Balancing Green Power
How to deal with variable energy sources

Balancing Green Power

How to deal with variable energy sources

David Elliott
Open University, UK

IOP Publishing, Bristol, UK

© IOP Publishing Ltd 2016

All rights reserved. No part of this publication may be reproduced, stored in a retrieval system or transmitted in any form or by any means, electronic, mechanical, photocopying, recording or otherwise, without the prior permission of the publisher, or as expressly permitted by law or under terms agreed with the appropriate rights organization. Multiple copying is permitted in accordance with the terms of licences issued by the Copyright Licensing Agency, the Copyright Clearance Centre and other reproduction rights organisations.

Permission to make use of IOP Publishing content other than as set out above may be sought at permissions@iop.org.

David Elliott has asserted his right to be identified as the author of this work in accordance with sections 77 and 78 of the Copyright, Designs and Patents Act 1988.

Media content for this book is available from http://iopscience.iop.org/book/978-0-7503-1103-8/page/about-the-book.

ISBN   978-0-7503-1230-1 (ebook)
ISBN   978-0-7503-1231-8 (print)
ISBN   978-0-7503-1232-5 (mobi)

DOI   10.1088/978-0-7503-1230-1

Version: 20160401

IOP Expanding Physics
ISSN 2053-2563 (online)
ISSN 2054-7315 (print)

British Library Cataloguing-in-Publication Data: A catalogue record for this book is available from the British Library.

Published by IOP Publishing, wholly owned by The Institute of Physics, London

IOP Publishing, Temple Circus, Temple Way, Bristol, BS1 6HG, UK

US Office: IOP Publishing, Inc., 190 North Independence Mall West, Suite 601, Philadelphia, PA 19106, USA

Contents

Preface		**vii**
Author biography		**viii**

1 Introduction: balancing variations — 1-1

1.1 Variable renewables — 1-1
1.2 Dealing with variable output–source correlations — 1-3
1.3 Balancing options — 1-6
1.4 The aims of this book — 1-8
1.5 A few words about terms — 1-10
   References — 1-12

2 The story so far: balancing with generation plants — 2-1

2.1 Grid balancing with variable renewables — 2-1
2.2 Balancing with fossil plants — 2-6
2.3 Non-fossil balancing plants — 2-9
   References — 2-13

3 The next challenge: energy storage — 3-1

3.1 Storing energy — 3-1
3.2 Battery storage — 3-3
3.3 Larger-scale storage — 3-7
3.4 Heat storage — 3-12
3.5 The way ahead — 3-15
   References — 3-15

4 Grid links to the future — 4-1

4.1 Electricity grids — 4-1
4.2 Supergrids — 4-3
4.3 Assessment of supergrids — 4-8
4.4 Local power — 4-12
4.5 Demand management and smart grids — 4-14
4.6 System choice — 4-17
   References — 4-18

5	**System integration**	**5-1**
5.1	System balancing options compared	5-1
5.2	System integration costs	5-3
5.3	Moving beyond LCOE	5-4
5.4	The ERP view	5-5
5.5	Other views on balancing needs and costs	5-7
5.6	The role of heat and balancing the mix	5-13
5.7	Balancing around the world	5-15
	References	5-19
6	**Making changes**	**6-1**
6.1	Renewables and grid balancing in the EU	6-1
6.2	The German approach	6-3
6.3	Flexibility, base-load and market design	6-6
6.4	Making the change globally	6-8
6.5	Institutional challenges	6-12
6.6	The challenges of change	6-16
	References	6-17
7	**Conclusion: all change**	**7-1**
7.1	The balancing challenge	7-1
7.2	Balancing technology issues	7-4
7.3	Balancing renewables without fossil fuel use	7-5
7.4	The challenges ahead	7-8
	References	7-10

Preface

The use of renewable energy is expanding rapidly, but some renewable sources are variable or intermittent. If the use of renewables is to expand further, ways have to be found to compensate for this variability. Fortunately, there are many, and as this book sets out to show, taken together, they can help balance grid systems as increasing amounts of renewable capacity are added. They include flexible generation plants, energy storage systems, smart grid demand management and supergrid imports and exports. This book outlines the options and explores how they might be integrated together in a reliable and sustainable energy system, avoiding wasteful curtailment of excess output and minimising the cost of grid balancing. It reviews a wide range of assessments of the viability of the various approaches, drawing on technical studies and strategic analysis from the UK, EU and USA. It also looks at how balancing issues are impacting elsewhere in the world, including in China. Written in an easy-to-access non-technical style, but with extensive references to more detailed studies, it argues that, with proper attention to balancing, flexible system development and energy saving, renewables can supply the bulk of the energy needed globally in the years ahead on a reliable basis.

Author biography

David Elliott

David Elliott BSc, PhD is Emeritus Professor of Technology Policy at the Open University, UK, where he developed courses and research on renewable energy innovation and development policy. He has published widely in the sustainable energy policy field, including an earlier IOP ebook on renewable energy.

IOP Publishing

Balancing Green Power

David Elliott

Chapter 1

Introduction: balancing variations

Renewable energy sources are large but some are variable. The use of variable sources for energy supply will require some way of compensating for this. Power grids already balance variations in conventional energy supply and variations in energy demand and the approaches used for this can be expanded to deal with renewables, even when renewables supply 20–30% or more of the electricity on a grid. The main option is to use fossil plants to back up renewables. However at higher levels of renewable input, heading for 50% and beyond, other, additional, balancing mechanisms may also be needed, including energy storage, smart grid demand response and supergrid imports and exports. This introductory chapter looks at the variations associated with the various renewable sources and then outlines the options for dealing with them, which are then explored in more detail in the rest of the book.

1.1 Variable renewables

Renewable energy sources currently supply around 23% of global electricity and there are prospects for this expanding dramatically, up to 50% by mid-century, and possibly much more (Elliott 2015). The basic resource is certainly very substantial. The Earth's natural flows of energy are very large. Incoming solar energy drives the winds, the waves and the hydrological cycle, while the gravitational pull of the Moon, combined with that of the Sun, creates tides (Boyle 2012). If these energy flows and sources can be harvested efficiently, they could provide more energy than mankind should ever need, for the indefinite future.

However, they vary. At any one location and time the energy available from these sources can be very low or very high. At times the energy flows can be overwhelming, as during storms, but at night solar energy income is effectively zero, and at some times of the year in some locations, wind speeds can also fall to zero. Humanity has learnt to live with these variations, but they are sometimes portrayed as a fundamental flaw in any attempt to rely on these sources for meeting our energy needs.

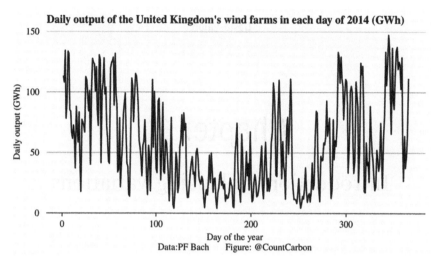

Figure 1.1. Daily wind farm output variations over a year in the UK (Carbon Counter 2015).

It seems so obvious that the wind does not blow all the time, and that the Sun does not always shine, that critics of the use of what are now called renewable energy sources find it easy to be dismissive. For them, a becalmed wind turbine or a solar array in the dark is a symbol of what they perceive to be costly, inefficient and unreliable technology, based on the use of intermittent sources.

It is certainly true that some renewable sources are variable. The winds can be very erratic, being dependent on complex weather systems, and so the availability of energy produced from wind turbines varies continually at any specific location, as figure 1.1 shows for the UK.

The range of the energy output fluctuations is wide for wind turbines, since their energy generation potential is proportional to the cube of the wind speed, so a small fall/rise can lead to a large change in output. The problems are different for solar: clouds are an issue and of course it gets dark at night. So it would seem that these sources are basically unreliable.

However, the reality is more complex. If you want to run a single wind turbine as your main source of energy you will certainly have problems, since its energy source is likely to be intermittent, or at least variable over a wide range. However, with many geographically dispersed turbines feeding energy to a grid system, the local individual variations in their energy input to the grid may not matter so much: if the wind speed is low in one location, it may be windy elsewhere. As later chapters will indicate, a geographic spread may not always give full compensation for the local variability of wind (it can be calm across wide areas), but it can usually smooth out the variations to some extent. Moreover, even a relatively limited local geographical spread of solar energy collectors can smooth out the very short-term local variations in energy generation due to individual clouds, assuming that it is not totally overcast over the whole area (Mills and Wiser 2010).

It is also worth pointing out that it is not just renewables like wind and solar that are variable. Most energy technologies have 'down-times' due to unexpected plant failures or scheduled maintenance. In the case of nuclear plants, they may also have to be shut down for several weeks every year or so for refueling. The result of this, along with the inevitable operational problems that can occur with these complex plants, is that, for example, the UK nuclear power plant fleet ran with an average annual load factor (a measure of the actual as opposed to theoretically possible energy output) of around 62% during the period 2007–12. Higher load factors have since been attained in the UK and also elsewhere, and it is claimed that new nuclear plants will be able to do even better, with load factors of 80–90% or more. However, wind plants are also improving. For example, in 2014 the UK's onshore wind plants recorded an annual load factor of just over 26%, and the offshore wind farms one of just over 38%, and this should rise as the technology improves and as offshore wind projects move further out to sea into areas where the wind regime is more reliable (DECC 2015).

Nevertheless, it is still clear that wind turbine load factors are relatively low. What that means in practice is that there will be periods when they are not running. Note that a 30% load factor does not mean they will only run 30% of the time: depending on location, they will typically run 70% of the time at some level of output. However, when they are not running, it is very obvious and possibly provocative to those who see wind energy as inefficient. By contrast, apart from changes in smoke or steam emissions, the actual level of activity is not so obvious with conventional fossil or nuclear-fired plants: what is happening inside cannot be seen. In fact, given that some of these plants are run with variable output to meet varying energy demand cycles, their annual load factors can be quite low; combined cycle gas turbine electricity generating plants in the UK recorded load factors of only 30.5% in 2014 (DECC 2015).

It is also worth noting that with large gigawatt-scaled nuclear and conventional plants, a breakdown or stoppage will lead to a large sudden reduction in energy availability. By contrast, with a large number of megawatt-scaled wind turbines on the grid, individual unit failure does not have much impact on the collective total output. The down-time for relatively simple wind projects is in any case usually much lower than for large and complex nuclear or coal plants: e.g. onshore wind plants are typically down for at most 2% of the year, compared to around 6% for scheduled maintenance and 6.5% for unscheduled stoppages for US coal plants (Jacobson and Delucchi 2011).

Even so, wind load factors are clearly not as good as those for nuclear plants, and most other generators can also deliver energy with higher overall reliability over the year. What that means, from the overall energy systems point of view, is that, in the absence of other measures, more wind and/or solar plant capacity will be needed to give the same total *annual* output as nuclear or fossil plants, and ways have to be found to compensate for the variations in availability within the year. That is the focus of this book

1.2 Dealing with variable output–source correlations

As illustrated for the case of wind in figure 1.1, the availability of electricity output varies in time at each location. There are also weather-dependent variations with some other renewables, notably solar energy, the potential availability of which also peaks at midday in each location, as the planet rotates.

The focus in the above has mainly been on wind, in part since, of the new (non-hydro) renewable electricity generating options, it has seen the largest deployment so far, with around 435 GW installed globally by the end of 2015.

However, the use of solar photovoltaics (PV) is accelerating rapidly, with nearly 200 GW now installed globally, despite it having a lower annual load factor of around 10–15%, depending on the location and technology. That weakness is to some extent offset by the ease of installation: PV arrays may soon become a common sight on many buildings, able to deliver electricity direct to users with no need for long-distance transmission. Moreover, this energy is often available at times when wind energy is low, for example in summer, when, in some locations, office daytime air-conditioning loads may be high, and will grow as climate change impacts. The variability of the solar input, and the low load factors, clearly present problems, but, as this example shows, there are some interesting locational and time-based correlations between the availability of this renewable source and demand. That is also true of other renewables, including wind, which is usually high during winter, when energy demand is high.

A key point is that demand for energy varies continually, with daily peaks and seasonal differences, and, since some of these variations are linked to the weather, there can be correlations with renewable energy availability.

It is worth exploring these and other types of correlations some more, since they set the scene for much of what follows.

The wind turbine load factors looked at above are based on the total national outputs from all the individual wind turbines in a country (the UK in this case) averaged out over the year. The outputs from individual wind turbines will obviously vary more than that from the combined set of wind farms across a large area, since, as noted above, there may be some geographical smoothing. It may be windy in some areas, but not others. It will also be windier at some times of the year than others. So at times the load factors for some turbines in some locations can be much higher (or lower) than the annual average.

Annual national load factors average out time-based variations like this, and also local site-based differences. Some turbines on good sites can have annual load factors of up to 50%, and at times much more, while a few will be sited in areas where wind speeds are such that average annual load factors are very low. An obvious question is, why are sites with low annual load factors used? The answer is that in some cases, and at some times—in winter, for example—their output may be well correlated with demand. So a low annual load factor may be acceptable.

Certainly it is true that, on average, winds are stronger in the winter, when demand for energy is usually highest. Moreover, while demand for heat is usually high in winter, when it is cold and winds are often high, there are also correlations between wind speed and energy demand, the so-called 'wind chill factor', which adds to the demand for heat. High wind flows can cool buildings down, thus raising energy demand, but they also lead to higher outputs from wind turbines. Wind chill can at times add 1 GW or more to UK heat demand, but if there is a large amount of wind generation capacity, the extra output from higher wind speeds may be sufficient to meet that higher demand, assuming that it can be met with electricity. This is a convenient and automatic correlation at the local and regional level, in terms of heat.

However, the seasonal correlations between energy supply and demand may not always be perfect. There can be long periods of cold weather (days, or sometimes weeks) with low or even no wind across wide areas. Moreover, the correlation with *electricity* demand is much less pronounced: it is less weather determined than heat demand (Sinden 2007). Even so, in general, wind availability patterns over the year do mostly dovetail quite well, in outline, with overall *energy* demand patterns.

There is also a good general inverse correlation between wind and solar energy availability. It is usually sunny in the summer, when winds are often low. As noted above, this may mean that there are also some helpful inverse correlations between the two sources in relation to local demand, although weather patterns are clearly complex and there can be cold, calm, dull days, or even weeks, over wide areas. So at times little of either will be available.

There are also more general geographical differences in wind and solar availability patterns. For example, wind flows across the UK are often westerly, as weather systems cross the Atlantic, so the UK (and Ireland) will experience peaks in wind energy availability some hours before the rest of northwestern Europe, while, unrelated to these weather patterns, the daily peaks in solar energy travel in the opposite direction, from east to west. Clearly the actual availability of energy from these sources at any one time and place will vary due to other factors, including local weather and topography. Interestingly, and perhaps surprisingly, US studies have found that, assuming a geographical spread of collectors, solar can be more reliable than wind. Depending on location, the level of predictive uncertainty, in relation to net solar generation output fluctuations across a region for an hour or more ahead, is generally lower than that for wind output (Mills and Wiser 2010).

As we shall see later, there has been much work done on these various local and regional correlations, looking especially at smoothing effects across wide geographical areas. Not all of the conclusions have been positive. For example, a 2011 study by Poyry consultants, looking at wind and solar correlations, concluded that 'in northern Europe the overall output of the renewable generation will be highly variable, and will not average out because of weather and geography' (Poyry 2011). However, as we shall see, other studies, including some over wider areas (e.g. pan-EU), have come to more positive conclusions: they will not be perfect, but there will be useful correlations. For example, one study of the impacts of cross-border EU interconnection concluded that 'wind and solar output is generally much less volatile at an aggregated level and extremely high and low values disappear', with interconnectors allowing local peaks and lulls to be smoothed and better matched to demand (Agora 2015). As we shall see, the inclusion of solar energy from North Africa in the resource assessment should improve the EU situation significantly. Larger-scale solar generation in desert areas offers some interesting correlations with demand if energy can be sent long distances to areas with colder/less sunny climates.

Some of the other renewable sources also offer interesting correlations. Waves are created by winds moving over long fetches of sea, and they can persist for some time after the wind has died down. So waves are in effect stored wind, offering, on average and at different times and locations, a slightly different set of correlations, although they are at a maximum in winter, when winds (and energy demands) are high. The tides are unrelated to solar or wind, and are regular, but cyclic, phased with the Moon, in a very predictable pattern at any one location, around the year. Some other renewables are not variable. Assuming they have collected a sufficient head of water, hydro-electric plants can supply energy continuously. Biomass can be collected and stored and used to generate energy continuously. Similarly, geo-thermal heat can be used to generate electricity and/or supply heat continually.

If a national energy system can use a mix of these various sources, the impact of the individual variability of some of them over the year, and also possibly over shorter timescales, may be significantly lessened. Moreover, for a complete national energy system, with many other generators also online, the impact of renewable variations is only one issue. What matters, for overall system reliability, is the net effect of these variations, along with those from other generators, as well as variations in demand. That is what the system has to deal with.

Subsequent chapters will look at how it might do that, but it is worth noting at this point that, as Sinden concluded from his early studies of renewable integration in the UK (including solar, wave and tidal energy, as well as that from wind), the use of multiple sources and multiple well-separated sites can reduce the need for back-up supplies (Sinden 2005)

Back-up can be provided in a number of ways. Grid systems already deal with variations in energy supply and demand. As we shall see, only when individual variable renewables like wind make up a large proportion of the energy input to a grid system, above around 20–30%, will there be a need for significantly increased balancing measures. Fortunately, there are many options. Box 1.1 presents a list of some of them. Following a short overview in section 1.4 below, subsequent chapters in this book will review these and other options.

1.3 Balancing options

The grid balancing options can be classed according to the point in the energy system where they operate. As chapter 2 indicates, at present, grid systems deal with variable supply (e.g. due to plant failures) and variable demand (due to consumers' energy use) mainly by ramping output from other plants up and down. Some plants are kept running at lower output ready for this purpose. They ramp up daily to meet the daily pattern of demand surges and other occasional extra demands on the system. They are sometimes called 'spinning reserve'. Others are kept on standby, offline, but ready to deal with major longer-term supply shortfalls or demand surges. In each case, the balancing is achieved by operating at the front end of the overall energy system, by feeding in extra (or less) energy.

The energy balancing in this system is in effect achieved by using stored fuel to make more electricity when needed. It is certainly easier to store fuel (coal, gas, oil,

> **Box 1.1. Some of the key balancing options.**
>
> In a seminal 2011 paper, US academics Mark A Delucchi and Mark Z Jacobson argued that there are at least seven ways to design and operate renewable energy systems using wind, water and solar sources (or WWS power, as they call it) so that they reliably satisfy electricity demand. Here is their summary.
> 1. Interconnect geographically dispersed naturally variable energy sources (e.g., wind, solar, wave and tidal).
> 2. Use complementary and non-variable energy sources, such as hydroelectric power, to fill temporary gaps between demand and wind or solar generation.
> 3. Use 'smart' demand–response management to shift flexible loads to better match the availability of WWS power.
> 4. Store electric power at the site of generation for later use.
> 5. Over-size WWS peak generation capacity to minimize the times when available WWS power is less than demand and provide spare power to produce hydrogen for flexible transportation and heat uses.
> 6. Store electric power in electric-vehicle batteries.
> 7. Forecast the weather to plan for energy supply needs better.
>
> (Jacobson and Delucchi 2011)

or biomass) than to store electricity once it is generated, so this *pre-generation storage* approach is likely to be the easiest way to deal with the larger variations that will occur when more renewables are used (Wilson *et al* 2014, Wilson 2015). Indeed, some say that this is all that will be needed, just more 'back-up' plants. For the moment, however, there is no need for *more* plants. Coupled with some other established balancing options, the existing range of spinning reserve and standby units is quite enough to deal with significant renewable variations. The conventional fossil fuel-fired plants just throttle back when renewable inputs are available and power-up more when they are less available. As more renewables are added to the grid, these plants will have to ramp up and down more often, lowering the carbon emission and fuel use reductions resulting from the use of renewables by small amounts. But the back-up plants already exist.

However, ideally, given their emissions, it would be desirable to get rid of these fossil-fired back-up plants. It may be possible to run some of them on storable renewable energy sources like biogas or green syngas, but it may also be necessary to think more about *post-generation storage*. That is actually often the first thing that is said when discussing variable renewables: they need storage to make them viable. The reality is more complex. Energy storage is expensive and there are many different types, with differing capacities, not all of them suited to dealing with the various types of variable renewables, or for long periods of supply shortfall.

At present, hydro-pumped storage is used for grid balancing and could be used more to balance variable renewables. It is very large in scale and suitable for bulk energy storage, but is obviously only available in hilly locations, where large reservoirs can be built. Grid links to these plants are of course possible, but it may be more convenient to have smaller storage facilities that can be built nearer to the variable generators and energy demand centers. As chapter 3 will show, there are some options at various scales, including advanced flow batteries, hydrogen and liquid air storage.

Storage is really about *time-shifting* energy from its generation to when it is needed. However, we can also try to shift *demand* in time: delaying demand peaks so that they coincide better with supply. This can be done by decoupling heavy energy-using devices from the grid at peak demand times. That might be achieved through the use of interactive smart grid systems or simply by variable energy pricing arrangements.

Another post-generation approach, which is looked at in chapter 4, is to shift energy over long distances from where there happens to be a temporary surplus to where there is a current local shortfall, i.e. shifting energy in time and *spatial* terms. The variations in renewable energy availability are usually defined by time and location; as noted above, wind fronts move across continents, so it may be windy in one place but not another: they can trade. To be done on a significant scale, this would require new, more efficient long-distance grid links, so-called supergrids. This opens up some large geopolitical issues, as we will see later.

So far we have been discussing electricity, storing it, delaying its use and shifting it elsewhere. But there are other forms of energy, most notably heat. It is easy to store heat and heat use is a large part of energy demand. Some renewables supply heat directly, while, in some other situations, it may be best to convert renewable electricity output into storable heat. It is also easy to store gas, as was noted earlier, as a pre-generation storage option. It may be that some post-generation electricity might also be best used to make gas to store to meet heat demand, or to generate electricity when and where needed. Gas can be transmitted over long distances with low losses. This opens up a whole new range of energy vector issues. For example, do we really need electricity as our main long-distance transmission and local end-use option?

A final set of balancing options involves the use of renewables. Not all of them are variable (hydro, biomass and geothermal can be considered as firm resources) and even those that are variable or cyclic (like wind, solar, tidal and wave energy) may have a degree of reliable output, often correlated with demand, as noted above. Reliability can be enhanced by better weather forecasting and by installing enough renewable capacity overall to meet demand, even when some plants are not able to deliver full output. As we shall see, there are problems with this approach: it will produce surplus output at times. However, some of that can be stored and used later to meet lulls in supply and demand peaks. The variability of renewables may thus, in effect, provide its own partial solution.

1.4 The aims of this book

As can be seen, there are many, often interacting, options for balancing variable renewables, although there is no one best choice, and, as this book tries to show, it is hard to decide on the right mix. However, what it also aims to show is that there are ways to manage variable renewables without excessive cost. It also brings into question whether, given the potential for decentralised and distributed energy generation and 'smarter', more efficient energy use, we actually need any large power plants, run continually in 'base-load' mode: they may be too inflexible to help with grid balancing. The advent of radically new energy supply technologies, perhaps unsurprisingly, may mean a need to develop radically new overall energy systems for balancing grids and using energy.

I have not gone into much technical detail regarding the various energy storage and balancing options. They are covered in outline in the first parts of this text, but, along with an introduction to the renewable supply options, I explored them in a little more detail in an earlier text in this series (Elliott 2013). The balancing options and issues are covered extensively in much more detail elsewhere (Boyle 2009, Apt and Jaramillo 2014, Jones 2014, Sorensen 2014).

Instead, the approach I have adopted here is an exploratory one, seeking to identify broader trends and issues related to grid balancing in general terms and in non-technical language. So it is not a textbook or technical primer. In terms of the specific focus of this text on grid balancing, there are also many more detailed high-quality studies, as noted above, although most would be hard for non-specialists to digest. With apologies to the specialists, I have tried to survey this rapidly expanding and quite complex field in simple terms, but with 'boxes' offering more detail or examples, and with extensive references to the more technical literature[1].

The focus of the book is mainly on the UK and to a lesser extent the wider EU and the USA, since that is where most work is currently being done, but the results and practices are of wider relevance and are already impacting on countries like China, which are pushing ahead with renewables rapidly. With some countries aiming for high renewable contributions (80% or more), and scenarios suggesting that most others could follow, the issues of grid balancing and system integration will clearly become increasingly important.

The main impetus for the adoption of renewables has been concerns over the likely social and economic impacts of climate change, coupled increasingly with concerns about the health impacts of air pollution. As the technology and markets have developed, costs have fallen to the point where many renewables are now competitive, in generation terms, with conventional sources. Although dealing with variability will add to the cost, the technologies for doing this are also developing rapidly, with costs falling. Moreover, given that they will enable the wider-scale use of renewables, long-term, they will help to avoid the growing social, environmental and economic costs of using fossil fuels and also the costs and risks of using nuclear power.

Most countries have only just started out on this journey, with balancing issues not yet becoming central, despite the rapid expansion of wind energy use. As the UK's Royal Academy of Engineering noted in a 2014 report on wind power, 'managing fluctuations in supply, including variable renewables such as wind power, is fundamental to the operation of the electricity grid', but 'to date, the balancing

---

[1] All the web links provided were accessed on 2 February 2016.

mechanisms already in place have been sufficient to cope with the amount of wind energy on the GB system' (RAE 2014).

However, that will change. This book provides a guide to what may be done in response. If you would like regular updates on subsequent technical and policy developments, one source is the weekly 'Renew your energy' blog post that I produce for the *Energy Research Web* run by the Institute of Physics[2]. This book draws heavily on my past postings on that.

In addition, the long-running bimonthly newsletter I produce, *Renew*, and linked posts, can be accessed online[3].

As will be seen, electrical engineering theory and practice underlies much of the field, and a good source for news and the discussion of developments related to new energy systems can be accessed via the website and the often very lively sub-conferences of the Claverton Energy Group, which draw on the expertise of a wide range of practitioners in the field[4].

The focus in this book is on technology, but clearly this does not exist in a social and political vacuum. The transition to using renewables as the basis of a sustainable energy system will involve social as well as technical choices and changes. I looked at some of the social and environmental implications and options for change in my previous book, *Green Energy Futures* (Elliott 2015), which set the context for some of the discussion on technical choices in the present volume. I should also note that, in part, this book grew out of a short commentary note I produced on renewable balancing for the inaugural issue of *Nature Energy* (Elliott 2016).

1.5 A few words about terms

I have tried to avoid specialist terms and units, and simply used watts and 1000× multiples, kilowatts (kW), megawatts (MW), gigawatts (GW), terawatts (TW) for the **power** of energy conversion systems, and the equivalent 1000× multiples of watt-hours for the **energy** supplied or used, kilowatt-hours (kWh) and so on, always remembering that, strictly, energy cannot be generated or consumed: it is just converted from one form to another with varying degrees of efficiency.

How and where it is converted and used (and transmitted or stored) is obviously a key issue for this book, and there are some problems with the terms commonly used to describe the emerging new systems for this. Perhaps the most obvious distinction is between large 'centralised' and smaller 'decentralised' power plants. However, 'decentralisation' is sometimes seen as implying small, local, off-grid generation, whereas it can in principle apply to any generator that feeds locally derived energy to users at the local level, including via local grids. A more precise term is 'distributed generation', implying that the feed in is at the local energy distribution grid level, not at or via the national transmission-level grid. That term is not always used, even by practitioners in the field. For example, two key organisations in the field are the Association for Decentralised Energy (based in the UK) and the World Alliance for

---

[2] http://blog.environmentalresearchweb.org/author/dxe.
[3] https://renewnatta.wordpress.com.
[4] www.claverton-energy.com.

Decentralized Energy (based in the USA), both of which are involved in promoting combined heat and power (CHP) systems, amongst other things. CHP plants can be quite large, serving whole communities. But CHP supplies heat and power locally, and technically, although some of the electricity may feed into the national grid, CHP is less 'centralised' than conventional plants. It may also be worth pointing out that, despite its name, domestic 'central heating' is not really centralized or indeed decentralised; this is just a term used to distinguish integrated heating units from home-heating systems using separate unlinked devices, although the energy used in either case is usually at present centrally supplied. While there are thus clearly issues, and there is room for confusion, the term 'decentralised', used in a pragmatic catch-all sense, may have to suffice to cover all the systems that use energy generated locally.

I have also used the term 'grid balancing' to cover all aspects of the process of responding to variable renewables, including the provision of 'back-up' capacity and grid integration measures (e.g. grid reinforcement). As will be seen, some analysts see balancing as just one aspect of a wider 'system integration' process. However, I feel all these aspects are best seen as *part of* the overall requirements for balancing, which is the end point aim. I have endeavoured to make clear which term is being used in specific cases.

Some technical terms are unavoidable. Electricity from power plants that can always be relied on to meet demand is labeled 'dispatchable', and if the plants are run continuously to meet the basic minimum demand level, they are called 'base-load' plants. When demand is low, and there is too much output from a power station, it may have to be 'curtailed', that is dumped, or the power plant throttled back. The term 'flexibility' is widely used in the grid-balancing context, to indicate how much supply or demand can be varied, for example, in response to variable renewable supply. Here is how it was defined in a 2011 report from the International Energy Agency: 'Flexibility expresses the extent to which a power system can modify electricity production or consumption in response to variability, expected or otherwise. In other words, it expresses the capability of a power system to maintain reliable supply in the face of rapid and large imbalances, whatever the cause' (IEA 2011).

Some specific terms are widely used in abbreviated form, as follows, and I have done the same, after their first initial full use in the text:
- CHP—combined heat and power (cogeneration of heat and power);
- CCS—carbon capture and storage (geological sequestration);
- LCOE—levelised cost of energy (cost averaged over plant life);
- PV—photovoltaic solar (solar cell electricity generation).

I have tried to avoid the use of the term 'power' to denote electricity, since this risks confusion, given that 'power' is often incorrectly used interchangeably with 'energy', but this, too, is often done, e.g. as in the case of the label 'CHP' and the 'power-to-gas' concept.

And finally, I have tended to use 'green energy' and 'renewables' interchangeably, although renewables are only one specific form of green energy, characterised by being naturally renewed. Other low carbon energy sources also exist and some may be viewed as being sustainable in the longer term.

Chapter summary

1. If the use of renewable energy is to expand there will be a need to deal with the variable availability of some of the sources.
2. Some renewables are more variable than others and there are some potentially helpful correlations between them and demand patterns.
3. Grid systems already deal with variable energy supply and demand patterns and can cope with those due to renewables up to moderate levels, mainly by varying the output from fossil plants.
4. At higher levels, additional balancing systems may be needed, but many of these already exist or are being developed, including storage systems, smart grid demand management and supergrids for long-distance import/export exchanges.

References

Agora 2015 The European power system in 2030—flexibility challenges and integration benefits *Fraunhofer Institute Report for Agora Energiewende* www.agora-energiewende.org/service/publikationen/publikation/pub-action/show/pub-title/the-european-power-system-in-2030-flexibility-challenges-and-integration-benefits/

Apt J and Jaramillo P (eds) 2014 *Variable Renewable Energy and the Electricity Grid* (London: Routledge)

Bassi S, Bowen A and Fankhauser S 2015 The case for and against onshore wind energy in the UK (Grantham Research Institute on Climate Change and the Environment, London School of Economics and Political Science/ Centre for Climate Change Economics and Policy, University of Leeds) www.lse.ac.uk/GranthamInstitute/wp-content/uploads/2014/03/PB-onshore-wind-energy-UK.pdf

Boyle G (ed) 2009 *Renewable Electricity and the Grid: the Challenge of Variability* (London: Earthscan)

Boyle G (ed) 2012 *Renewable Energy* (Oxford: Oxford University Press)

Carbon Counter 2015 UK wind turbine generation as rendered by Carbon Counter (https://twitter.com/countcarbon), based on data from www.elexonportal.co.uk

DECC 2015 Digest of UK Energy Statistics (DUKES) (London: UK Department of Energy and Climate Change) www.gov.uk/government/uploads/system/uploads/attachment_data/file/447647/DUKES_2015.pdf

Elliott D 2013 *Renewables: A Review of Sustainable Energy Supply Options* (Bristol: Institute of Physics Publishing)

Elliott D 2015 *Green Energy Futures* (London: Palgrave Pivot)

Elliott D 2016 A balancing act for renewables *Nature Energy* **1** 15003 www.nature.com/articles/nenergy20153

IEA 2011 *Harnessing Variable Renewables* (Paris: International Energy Agency) www.iea.org/publications/freepublications/publication/harnessing-variable-renewables.html

Jacobson M and Delucchi M 2011 Providing all global energy with wind, water, and solar power. Part II: reliability, system and transmission costs, and policies *Energy Policy* **39** 1170–90

Jones L (ed) 2014 *Renewable Energy Integration* (London: Elsevier)

Mills A and Wiser R 2010 *Implications of Wide-Area Geographic Diversity for Short-Term Variability of Solar Power* (Berkeley, CA: Lawrence Berkeley National Laboratory) http://eetd.lbl.gov/ea/emp/reports/lbnl-2855e.pdf

Poyry 2011 *The Challenges of Intermittency in North West European Power Markets* (Oxford: Poyry) www.poyry.com/news-events/news/groundbreaking-study-impact-wind-and-solar-generation-electricity-markets-north

RAE 2014 *Wind Energy—Implications of Large Scale Deployment* (London: Royal Academy of Engineering) www.raeng.org.uk/windreport

Sinden G 2005 *Renewable Electricity Generation and Electricity Storage* (Presentation, Cambridge University, Energy Storage seminar, June 2005) www.cambridgeenergy.com/archive/2005-06-22/CEF-Sinden.pdf

Sinden G 2007 Characteristics of the UK wind resource: long-term patterns and relationship to electricity demand *Energy Policy* **35** 112–27

Sorensen B 2014 *Energy Intermittency* (London: CRC)

Wilson G, Rennie A and Hall P 2014 Great Britain's energy vectors and transmission level energy storage *Energy Procedia* **62** 619–28

Wilson G 2015 Tesla batteries might power your home but stored fuels will still run the country *The Conversation* **2** July 2015 https://theconversation.com/tesla-batteries-might-power-your-home-but-stored-fuels-will-still-run-the-country-42859

IOP Publishing

Balancing Green Power

David Elliott

Chapter 2

The story so far: balancing with generation plants

Fossil fuel-fired power plants are used to balance variations in grid energy supply and variations in demand, by ramping their output up and down. Some types of modern gas-fired plants can do this rapidly and with relatively low efficiency losses. With variable renewables on the grid, they would have to do this more often, imposing small extra cost and emissions penalties. The latter can be avoided by using 'green' energy sources to fuel the gas plants, but there may be limits to how much green gas will be available. In principle gas plants can provide balancing for both short-term variations and long-term lulls, using stored fuel. Nuclear plants may not be able to balance short-term variations, but some might play a limited role in dealing with long lulls. Some types of renewable are less variable and can play a role in short and possibly also longer term balancing, and wind does have a degree of statistical reliability. However high-reliability fossil plant backup remains the most popular and cheapest balancing option at present.

2.1 Grid balancing with variable renewables

When renewables were first introduced on a significant scale from the 1990s onwards, initially in the industrialised countries, there were worries that they would disrupt power grids and that there would be a growing need for 'back-up' plants to cope when wind and solar plants were not producing electricity. These concerns have proved unfounded so far, and a simple description of how grid systems work indicates why.

When the first few megawatts of wind turbine capacity are added to the grid system, nothing much changes. When the wind turbines run, their electricity feeds into the power grid and, if there is too much electricity being generated overall, the grid controllers throttle back on output from other plants to compensate. It is the same for other renewables, like PV solar. When and if there is less wind or solar electricity available, the conventional plants are ramped back up. Mostly it will be fossil-fired plants that are used for this balancing, not nuclear plants, since the latter cannot vary their output easily and quickly. The result of the ramp downs is that there is a reduction in the amount of carbon dioxide produced. So these renewables are operated in a fuel- and carbon-saving mode. On that basis, every kWh of renewable electricity displaces a kWh of fossil generation, with no disruption and no need for extra back-up plants: they already exist, although as they age, newer, more efficient and flexible versions can replace them.

That, at least, is the simplified picture. In reality it is a little more complicated. Assuming limited renewable capacity on the grid, the variations introduced by the wind/solar inputs will be small and will usually be lost in the 'noise' of other variations on the grid, due to output changes from all the other plants and the often large demand changes. It is the net aggregated changes in energy demand and available grid-linked electricity, along with changes in voltage and frequency, that the system operators have to deal with, and they select the cheapest and easiest ways to deal with them. See box 2.1 for a simplified guide to the UK system.

Box 2.1. Grid balancing—how it is done now.

The national power grid system in the UK maintains voltage and frequency stability within closely defined limits (240 V and 50 Hz), although the small variations permitted do mean that the energy delivered at any point in time can be allowed to change slightly without having to adjust supply significantly. An online picture of the process of frequency adjustment in the UK as it happens is available[1].

When and if some supply plants fail or demand rises more substantially, grid-linked power plant output is increased to compensate, and if there is a continuing shortfall, extra capacity may be brought online, some of it fast start up so-called 'peaking' plant, or standby capacity. Pumped hydro reservoir storage capacity may also be used. This can deliver output very rapidly. Or electricity may be imported via interconnectors from mainland Europe. It depends on what is cheapest at the time. In the next level of system protection, rarely used, there are options for bringing extra, old or ancillary plants online for a while, and for cutting loads temporarily: some companies contract for electricity on an 'interruptible supply' basis and pay less for the energy.

Given that high prices can be charged for electricity when there are sudden shortfalls, some grid supply companies specialize in meeting the occasional supply shortfall/peak load events by contracting with small generators who may have surplus. For example, Flexitricty, based in Edinburgh, can call on a range of small plants of various types contracted as part of the short-term operating reserve (STOR). It works well. As Flexitricity has noted, on one occasion in 2009, when a nuclear plant failed at peak demand time, the STOR system fired up and within five minutes filled the gap for an hour or so until demand fell (Martin 2013).

Smaller shorter-term power losses, of the sort that might often occur with wind variations, can easily be dealt with by frequency adjustments and/or by using STOR.

---

[1] www.dynamicdemand.co.uk/grid.htm.

> If there are longer lulls, several hours or even days, then other balancing measures are needed, but as is discussed below, there are many options, including ramping up all the fossil plants to maximum output, and using the strategic reserve capacity fully. As we shall see later, the UK has now contracted, with around 50 GW of mostly existing capacity to be available to meet shortfalls. Given that the total UK electricity generating capacity is around 70 GW, that is a large proportion of it: basically, the generation system is used to balance its own variations.
>
> With variable renewables on the grid, the rest of the system continues to provide balancing, although, in addition to the variable energy outputs, with renewables, there are also slight differences in terms of contributions to system stability. For example, large conventional power plants, with their large, heavy rotating turbines driven by high temperature steam/gas pressure, when coupled to the grid, provide an inertial buffer against frequency shifts. The engineers talk in terms of 'synchronizing torque'. Most renewable-driven devices (wind, wave and tidal turbines) do not have so much in-built rotational inertia, and solar has none. However, the grid system operators are able to compensate for this, as part of the process of balancing variable outputs and demand, although the loss of large synchronous generators means that more frequency response systems are needed to maintain system stability. This of course assumes that it is vital to retain a high level of frequency stability, which may not be the case, at least on the demand side, given that, nowadays, few consumer devices rely on tightly controlled and synchronised mains frequency.
>
> If you want to see the UK energy mix in action, with the daily, weekly monthly and annual shares of the various sources shown graphically, see the Gridwatch website[2], which also has French data. For Germany, see online[3] and the regular reports from the Fraunhofer Institute, which include full breakdowns of the supply mix in action, helpfully summarized and continually updated[4] (Fraunhofer 2014).

Although, as box 2.1 indicates, there are a range of options, depending on the scale and duration of the disruption, most likely it will be small- to medium-sized gas-fired plants that are used for short-term grid balancing, since they are usually much more flexible than large coal plants. Some of the latter can take up to an hour to change their output by 50%, while some gas plants can do that and more in minutes, especially simple 'open cycle' gas turbines. The more efficient combined cycle gas turbines are somewhat less flexible (see table 2.1).

There is, however, a problem. Operating gas turbine plants flexibly can reduce their efficiency and reliability. Ramping them up and down regularly adds to wear and tear stress, and operating at below optimum levels can worsen their fuel-use efficiency, so that they produce more carbon dioxide gas per unit of fuel used. In the early days of renewable expansion it was sometimes argued by critics that this meant that the carbon savings from using renewables would be undermined, or even wiped

---

[2] www.gridwatch.templar.co.uk.
[3] www.agora-energiewende.de/en/topics/-agothem-/Produkt/produkt/76/Agorameter.
[4] https://energy-charts.de/downloads.htm#.

Table 2.1. Ramp up rates for fossil fuel plants (NEA 2012).

	Start-up time	Maximal change in 30 s	Maximum ramp rate (% min^{-1})
Open cycle gas turbine	10–20 min	20–30%	20
Combined cycle gas turbine	30–60 min	10–20%	5–10
Coal plant	1–10 h	5–10%	1–5

out entirely. In reality, these savings are very much larger than any losses due to variable back-up plant operation. It does not take much thought to see why. All other things being equal, if the renewable input did not exist, the fossil plants would be generating emissions. With the renewable input, a proportion of those emissions, on a kWh per kWh generation-replaced basis, is avoided, since the fossil plants operate at lower output, using less fuel. Even if their fuel efficiency is low at that output level, the marginal extra emissions per kWh will be small, much smaller than the savings. Here is how a report from a UK House of Lords Select Committee put it in 2008: 'The need to part-load conventional plant to balance the fluctuations in wind output does not have a significant impact on the net carbon savings from wind generation'. (House of Lords 2008)

However, the effects are not negligible and, as more renewables come on the grid, there will be more variable use of fossil capacity. So, efforts have been made to develop balancing gas plants that can vary output with lower efficiency losses and, crucially, do so more rapidly. Although wind availability prediction techniques have improved significantly, it is still helpful to be able to ramp output from gas plants up and down fast, since there is a 'double whammy' problem: the gas turbines need to be able to ramp up fast when demand is rising and renewable input is simultaneously falling, and also ramp down fast when demand is falling and renewable input is rising. Some can do this quite well. For example, GE's 510 MW *FlexEfficiency50* is claimed to be able to ramp up or down at a rate of 51 MW min^{-1} (i.e. 10% min^{-1}) and still deliver 61% energy conversion efficiency, while Siemens and Alstrom have gas turbines that can ramp even faster, with low losses (Probert 2011). Moreover, this is not a static field. As renewables have expanded, so has the incentive for gas turbine designers to develop even better turbines to cope with the increased cycling requirements (Bade 2015).

With balancing options like this being available or emerging, early studies suggested that it would be possible to cope with renewable inputs of up to around 20% of the total on the grid without significant problems (Grubb 1991, Boyle 2007). In the event, this estimate has been overtaken by developments in pioneering countries like Denmark, which now gets over 40% of its annual electricity from wind, and is heading for 50%. How does Denmark balance that? Fossil plant ramping will be part of the answer, but at some times of the year, and at night time especially, Denmark's wind capacity generates all the electricity that is needed and maybe more, even with all the fossil plants ramped down to minimum. So the surplus is exported to Norway, where it can be used to pump water uphill into some of that country's extensive hydro reservoirs. When electricity is needed, this extra head of water is run out through the hydro turbines. If Denmark needs it at some point, for example when its wind generation is low, its stored wind energy can in effect be re-imported.

What this example indicates is that it may be *excess* renewable output that is the issue, not *shortfalls*. Surplus generation is likely since, given that renewable output can vary, in order to help meet average demand, or even perhaps peak demand, more capacity may be installed than would be necessary if it was not variable. This, in fact, is a form of balancing strategy—an *overcapacity* approach. Extra renewable capacity is installed to try to ensure that demand can be met even when the renewable input is low and demand is high. However, this has an obvious disadvantage. The resultant occasional surplus production when demand is low and renewable or other supply is high may be too much to handle and so require what is called 'curtailment', the involuntary reduction of output to the grid from the renewable generator.

While it is clear that there will be a problem generally if supply exceeds demand, there can be a range of specific reasons for curtailment, even when the total renewable capacity on the grid is low. For example, in some countries, local grids are weak and may not always be able to deal with the full output of a local wind farm as well as the output from other local conventional plants on the grid, if that is also high. If the latter cannot be ramped down, the wind plant output has to be halted or dumped. It is a grid *congestion* problem. The same can happen when there is surplus available from a renewable generator, even if the grid system is robust, if there are large inflexible conventional plants on the grid, for example a major nuclear plant.

Full curtailment of local surplus output from renewables may not be the only option, since, if the grid system can carry the excess, some of it may be sold off at low prices, possibly below retail or even wholesale prices. From the generators' point of view, this can make sense, since the marginal generation costs with renewables like wind and solar are very low (there are no fuel costs), but from the overall system's point of view, dealing with, in effect, negative prices can be problematic. Dumping cheap energy like this can undercut the viability of the other plants. As we shall see later, in some countries, notably Germany, low marginal cost renewables like PV solar have made it hard for gas plants to compete in the peak demand market, since demand peaks in the middle of the day, when there is often plenty of solar output. So gas makes way for solar (and wind) at that point and avoids the need for curtailment.

Although curtailment might be seen as a way of balancing the grid, it is obvious wasteful. Valuable green energy is being lost. The scale of this loss may be relatively small for the moment, given the relatively low level of renewable input. An early 2011 estimate by the UK National Grid suggested that, with small wind inputs, there would only be three days a year when high UK wind output coincided with low demand for it. A 2014 US study found that wind curtailment, mainly in that case due to weak grids and grid congestion, only cut 4% or less of annual wind plant output in areas where it occurred (Bird *et al* 2014).

However, the renewable input has now reached over 20% in the UK and the curtailment problem there and elsewhere will grow as renewables expand further, unless other measures are adopted. A study by the UK Energy Research Partnership claimed that in a hypothetical 100% UK renewables scenario, even with some balancing measures, around 8% of annual renewable generation would be curtailed (ERP 2015). And some US studies have put the possible scale of curtailment, if no flexible balancing is available, much higher, at up to 17.8% of total annual renewable output in one study with a large PV solar input (Mills *et al* 2013).

Apart from being wasteful, curtailment is also very provocative, since, in some countries, contractual protection is negotiated to compensate generators when and if their output is not used. In the UK this led to media headlines about 'wasted wind-power millions', although it has to be pointed out that many conventional generators, not just wind farms, also enjoy these so called 'constraint payments'. The UK's National Grid company indicated that in 2012/13 wind projects only received 7 million UK pounds out of total national constraint payment for all projects of 170 million UK pounds. To be fair, the proportion of the payments received by wind projects has to be put in context: at that time, wind projects only made a relatively small energy contribution compared to the rest (about 5%). However, set against that, the proportion of the payments to wind plants had actually gone down from 2011/12, when the total was 324 million UK pounds, with 31 million for wind constraints. The reduction to 7 million UK pounds in 2012/13 was due to grid upgrade investments, and was despite an increase in the number of wind projects (NG 2016).

Clearly, upgrading grids will help, but as renewables expand, the curtailment problem will grow. As we have seen, while local grid constraints can play a role, fundamentally, curtailment is a result of the variability of renewables and the need to compensate by having more capacity installed than is needed most of the time. However, the problem of surplus production may actually be turned into a partial solution to the balancing problem: store it for later use to meet shortfalls. As we shall see later, this can be done in a number of ways. For example, in addition to a range of other storage options, including pumped hydro storage, surplus electricity can be used to make syngas (e.g. hydrogen produced by electrolysis) for storage and then used for power generation when needed. We will be looking at this so-called 'power-to-gas' idea later. There is also the option of exporting excesses to countries where there is currently a shortfall, or storage capacity: we will also be looking at the 'supergrid' idea later.

2.2 Balancing with fossil plants

These and other balancing options, including various other forms of post-generation green electricity and heat storage, could make it possible to deal with much higher renewable inputs than are used at present, while limiting curtailment and the need for excessive renewable capacity. However, for the moment, the use of existing fossil plants remains the easiest option for grid balancing. In addition to the use of standard grid-linked gas turbines, other balancing options include the use of a range of other fossil plants at various scales to provide back-up reserve power to cover short- or long-term shortfalls. As illustrated in table 2.1, the plants on the grid all have differing degrees of fast 'ramp up' capacity, but most of the options can play key roles for short- or longer-term balancing, as is explained in more detail in box 2.2.

> **Box 2.2. Short- and long-term reserve capacity.**
>
> The common, standard, two-phase combined cycle gas turbines (CCGTs) have high generation efficiency but are not as good at flexible operation as the simpler open cycle gas plants (OCGTs), some of which are so-called 'aeroderivatives', based on jet engine designs. Typically, OCGTs can ramp up to 100% of full output in minutes (table 2.1). So they are good for fast-response short-term balancing, meeting sudden demand peaks or supply shortfalls. However, they have lower efficiency than CCGTs. So they are not ideal for longer-term balancing.
>
> There are also many small reciprocating car engine-type plants designated to run to meet sudden shortfalls. These include some off-grid diesel-fired standby/emergency generators, which can start up from cold almost instantly. Diesel is dirty in emission terms (though not as dirty as coal), but since these plants only operate very infrequently and for short periods, that is less of a worry. Moreover, some could run on green diesel, or other low carbon biofuels. There is around 20 GW of standby capacity like this around the UK, often in institutional settings and designed for emergency power. Some could be grid linked and used for short-term balancing, joining the 8 GW or so of other plants in the UK's STOR system.
>
> In addition, the UK, like most countries with large grids, has some large-scale reserve capacity, offline plants that can be brought into use to meet occasional longer-term shortfalls. Most may be old low-efficiency fossil-fired (gas, coal or even oil) plants, but as with the STOR fossil/diesel plants, that may not matter too much, since they would not be used often. They can be used to meet demand when there are long lulls in wind or solar availability. Germany has just put some of its old lignite-fired plants into this offline reserve category, mothballed but potentially available. As noted in box 2.1, in the UK, a special 'capacity market' has been set up to provide support for the retention of a range of plants, around 50 GW in all, that can provide balancing services, most of the contracted capacity being old existing conventional plants, including some diesel STOR units. Note that this capacity market is not just for balancing renewables, it is mainly for dealing with other possible shortfalls and maintaining a plant margin to ensure energy security around the year.
>
> How much balancing and reserve capacity will be needed in future? As the proportion of renewables grows, and old back-up plants retire, extra short- and long-term balancing capacity will be needed. However, this new capacity does not all have to be conventional generation plant. As mentioned in chapter 1 and explored in more detail in subsequent chapters, there are several other balancing options, including storage, smart grid demand-side management and supergrid interconnecters. The initial UK capacity market auction process did include some of these options, and similar developments are occurring elsewhere, so in future these new options may well begin to dominate. Nevertheless, there may still be a need for conventional reserve plant for some while. This is an issue that is looked at in chapters 5 and 7.

While, as box 2.2 makes clear, some gas plants can be good at fast short-term balancing, some of them, and some other fossil plants, can also play a role in coping with long-term shortfalls. Figure 2.1 provides an overview of the relative flexibility levels of UK power plants, indicating how long it takes them to reach full output from initial low levels. Once they have done so, however, they can all meet demand

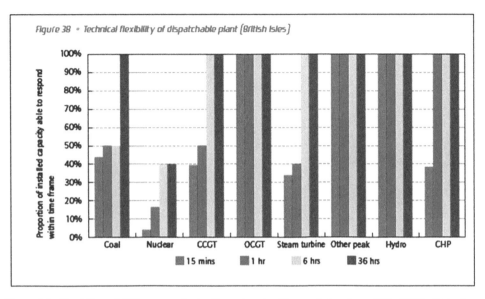

Figure 2.1. Flexibility of UK power plants. Reproduced with permission from IEA (2011). Data from OECD/IEA.

continually. Whereas most post-generation storage/balancing systems can only balance shortfalls for a few hours (e.g. batteries) or for up to a day or so (e.g. pumped hydro), the use of the existing fossil plants can deal with long periods of low renewable availability using easily stored fossil fuel, stored for use in pre-generation mode. This is already done on a very large scale and some say it is the best bet, at least for the near future (Wilson 2015). Certainly pre-generation storage of fossil fuels represents by far the largest energy storage option used at present. Nuclear represents a special case. As noted earlier, and as figure 2.1 shows, it is inflexible, with poor ramping ability, while hydro is the opposite, offering almost instant full start-up and a high degree of flexibility. We will look at these two non-fossil options in the next section.

In terms of the fossil fuel options, coal and oil are stored in bulk. These sources are unlikely to be acceptable environmentally as major players in the future, but gas may be. There is around 40 000 GWh in UK gas store capacity, but the gas grid itself also offers a pre-generation storage option: 'linepack' storage in the UK gas transmission pipeline has a capacity of around 5000 GWh (Wilson *et al* 2014). So the UK gas storage potential is quite large, as is the associated pipe network. That is not surprising, given that the UK gas system handles between three–four times more energy than the electricity system. Interestingly, it has lower distribution losses than the electricity grid and is able to deal with the much larger swings in energy use due to the typically large daily and seasonal heat demand variations (figure 2.2). Storable gas can and does take the strain of variable demand on a much larger scale than electricity.

While some countries have even larger gas stores, not all countries have this gas option, at present, on as large a scale. As we shall see later, that could change, for example as green gas options open up, and it may be that, in some contexts, gas

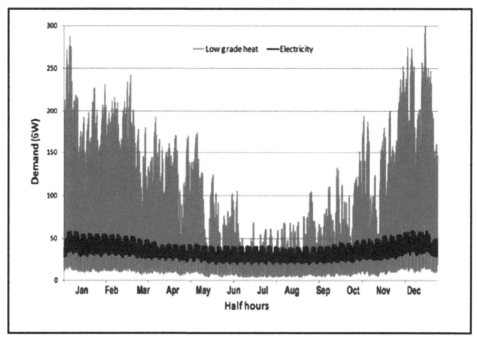

Figure 2.2. UK electricity and low-grade heat: half-hourly demand pattern in 2010. Reproduced with permission from Douglas (2015). Data from Samson and Strbac (2012).

represents a better, more flexible, way forward than the use of electricity for transmission, storage and balancing. However, whichever transmission and storage approach is used (gas or electricity), for the moment we are still talking here about using fossil fuels as the energy source. Given that there are emission problems with burning fossil fuels, that is not ideal. Fortunately, there are other possible back-up approaches to balancing.

2.3 Non-fossil balancing plants

The use of fossil plants for balancing produces extra emissions and although, as we have seen, they are relatively marginal at present, when and if renewables expand, they could grow. In theory, these extra balancing-related emissions could be collected, along with the emissions produced the rest of the time the fossil plant runs. However, so far, carbon capture and storage (CCS) systems have not developed on any significant scale. Even if they do develop successfully, they will be costly, and, it is arguably rather odd to try, in effect, to balance variable renewables by storing ever-accumulating amounts of carbon dioxide gas somewhere, keeping it secure forever. Moreover, the addition of CCS may make it harder to ramp fossil plants up and down to balance variable renewables, since their overall efficiency may suffer.

So the use of low-carbon green gas and syngas, or in some cases solid biomass, for fuelling balancing plants has attractions. How far it will be possible to go in terms of using biogas for generation and balancing remains unclear: biogas resources may be limited. For solid biomass, there are land-use constraints to energy crop growing, although that will not apply to the use of biomass wastes, since they already exist. The prospects for the production of biogas using anaerobic digestion (AD) of farm and food waste are quite good, with about 18% of UK gas possibly being AD biogas by 2020. However, most of that should arguably be used for heating, not electricity generation. Moreover, some of the wastes ought to be avoided in the first place. There are other potential sources of green gas, including various synthetic gas production options and the emerging power-to-gas surplus renewable electricity conversion option, although that is as yet undeveloped. Even so, the green gas option may play a significant role in generation and balancing in the future (Abbess 2015).

It might be thought that nuclear power plants could also play a balancing role. After all, nuclear reactors do not produce carbon dioxide directly, even if making the fuel for them does. However, they are inflexible and usually run 24/7 to recoup their large capital costs. Their output can be varied relatively slowly (figure 2.1), for example to meet the standard daily demand cycles, as is done in France. So they might be able to ramp up to help cover the occasional long periods of low renewable availability, although that would mean they would have to be run at less than full power most of the rest of the time, which would worsen their economics. Some old plants might be used like this (Cany *et al* 2016), but few would suggest building new nuclear plants just for this purpose, since there are many cheaper and more flexible options. Certainly nuclear plants, new or old, cannot easily and safely ramp up and down rapidly and regularly in response to the frequent short-term variations in renewable output. It takes time for the radioactive xenon gas produced when activity levels fall to disperse. This can disrupt safe plant operation.

The Nuclear Energy Agency quotes a ramp-up rate range of 1–5% of full output per minute, similar to coal plants (NEA 2012), but this may be optimistic. For example, it has been reported that some current French plants need 30 min to ramp up or down between 60% and 100% of their full output and one hour to ramp up and down between 30% and 100% of full output, while longer times have been cited for some other plants (IEA 2011). EDF have said that the new, as yet untested, European pressurised water reactor 'can ramp up at 5% of its maximum output per minute, but this is from 25% to 100% capacity and is limited to a maximum of two cycles per day and 100 cycles a year. Higher levels of cycling are possible but this is limited to 60–100% of capacity' (EDF 2008). So it would be of little help in balancing rapid changes in renewable output on a regular hour by hour basis.

As we shall see, there are many much more flexible options for balancing, in addition to the fast ramp-up gas plants looked at in this chapter. It is also worth noting that the balancing issue may be less of a problem than it initially seems. Although they are variable, the use of renewables like wind and solar can offer some level of reliable or 'firm' output. There have been many attempts to come up with estimates of how much wind capacity can be considered to be firm. One rough rule of thumb approach suggested for wind projects is that the so-called 'capacity credit', the capacity that can be relied on to meet demand with the same level of certainty as other plants, will be the square root of the total wind plant capacity (suitably expressed numerically: the square root of 1 is 1!). So if the plant capacity is 25 GW, the capacity credit will be 5 GW. That does not mean it will be 100% reliable. No power plant can be 100% reliable. However, statistically, wind projects can be expected to provide a minimum collective output with a high degree of reliability.

On that basis, some fossil plant could actually be replaced, while still maintaining the same level of system reliability, with wind plant adding to the capacity margin, the level of capacity over and above that needed to meet peak demand. As one early UK study put it, 'for the level of wind penetration of 26 GW, about 5 GW of conventional capacity could be displaced, giving a capacity credit of about 20%'. (Dale *et al* 2004)

In 2014, the Royal Academy of Engineering (RAE) noted that Ofgem, the UK gas and electricity market regulator, 'had determined that 17–24% of wind capacity could be counted towards the overall margin'. However, the RAE warned that 'this does not mean that wind is expected to produce at least 17% or more of its total installed capacity all the time; this measure is part of a more general probabilistic calculation on the overall risk that supply might fall below demand' (RAE 2014).

The terminology can get a little confusing in this area, given that some studies refer to load factors (which were looked at in chapter 1) as 'capacity factors', sometimes with a slightly different definition. I have stuck with the term 'load factor' to avoid confusion with 'capacity credits' and 'capacity margins'. To reiterate, the 'capacity margin' indicates how much 'head room' the system has above peak demand. The 'capacity credit' is the percentage of the wind capacity that can be relied on statistically to meet demand. But to add to the complexity, Ofgem uses what it calls the 'equivalent firm capacity', which it defines as the quantity of firm (i.e. always available) capacity required to replace the wind generation in the system so as to give the same level of security of supply. Its 2014 assessment looked at a range of scenarios, each leading to different equivalent firm capacities. Depending on the scenario (and demand levels) and the year (running up to 2018/19), they range between 14.8% and 26.2% of the total installed wind capacity (Ofgem 2014). Given that by late 2015 the UK had around 13.5 GW of wind capacity, that means it offered between about 2.0 and 3.5 GW of 'equivalent firm capacity'. A report published by the Adam Smith Institute, not known for its support of wind energy, puts the capacity credit at 2.3 GW for a UK system with 10 GW of wind capacity, which thus falls roughly in line with the Ofgem figures (Aris 2014).

To reiterate, whatever the terms used, this does not mean that that amount of capacity will always be functioning 100%, it just reflects how much of the total wind capacity can be relied on statistically to give the same security of supply as other, firmer sources. Note also that the capacity credit will reduce as the proportion of wind in the same general location increases, since there would then be a larger chance of low totals. Even so, some fossil plant could in theory still be retired without incurring significant energy security risks.

However, that is contentious. Certainly, estimating capacity credits for speculative systems in the future is no easy matter and the statistical projections approach has been challenged as unreliable. Critics insist that wind cannot replace much, if any, conventional capacity. As Professor Michael Loughton put it in an early study, 'the capacity credit will never be more than the planning margin', so that 'the total conventional plant capacity will never be less than the peak load irrespective of the amount of added wind capacity'. In which case, wind and other variable renewables 'can replace energy supplied from conventional sources, but not the need for most of their capacity' (Loughton 2007).

This formulation has resurfaced over the years in various versions, and does of course allow for some *marginal* capacity credit replacement, but, as we shall see, in the more extreme versions, it is suggested that the capacity credit is negligible, or even zero, and that back-up capacity has to be provided 100% for all variable renewables: they cannot replace any conventional plants, due to their unreliability.

Clearly, not everyone agrees. Variability is certainly a constraint, but no plants can actually deliver 100% firm power. As the Royal Academy of Engineering study mentioned above noted, the power system runs on the basis on probabilities and back-up, and that is true even for nuclear plants: they too have back-up costs. As noted in chapter 1, they can go off-line unexpectedly and the rest of the system has to be available to back them up.

Wind and solar may have lower levels of reliability in system terms but, statistically, given an integrated and dispersed system, there may be a low probability of zero collective availability. So back-up requirements may be less. Moreover, if other balancing measures are available, the fossil back-up requirement would be reduced, as we will see later. It would reduce even more if several different types of renewable source were run together on the grid, given that, as illustrated in chapter 1, they can have differing variation patterns and potential correlations with demand.

It is not easy to model these complex multi-source temporal interactions, and we will be coming back to look at some of the issues later, in terms of how much fossil-back-up might be needed, but on the basis of his early work on wind, wave and tidal power, Sinden at Oxford University asserted that 'the maximum additional back-up required because of intermittent electricity generation is never greater than the equivalent amount of conventional generating capacity being displaced' (Sinden 2005). Moreover, as we shall see, it could be less.

All that said, the availability of electricity from non-variable renewable sources like hydro, geothermal and storable biomass, would offer much higher reliability. Plants using these sources can be treated as firm capacity and can be run flexibly. Biogas-fired plants are similar to gas plants, with good ramping capacity, while geothermal plants run in combined heat and power mode, as with fossil or biogas-fired CHP units, can vary their heat and power output ratios to balance varying needs. As figure 2.1 indicated, hydro has very high ramp-up abilities and is very flexible, which is why it is already widely used for balancing, both with and without pumped storage. There are also tidal energy options, which, although not firm, can have high reliability (the tides are very predictable), especially if use is made of barrages or tidal lagoons for pumped storage, in which case their hydro-type flexibility could be further enhanced. In the case of the UK, and possibly other locations, a geographically distributed network of tidal current turbines, barrages and tidal lagoons around the coast could also take advantage of the fact that high tide occurs at different times at each point, enabling the network as a whole to deliver more continuous output, especially if the storage potential of barrages and lagoons was also used (Yates *et al* 2013, Elliott 2015).

However, even with these extra firm, or near firm, renewable supply inputs added longer term, and possible correlations in availability and demand matching, as looked at briefly in chapter 1, it will be necessary to make more use of other balancing options. While grid-linked fossil plants may have the edge at present, energy storage systems are often seen as an alternative, or at least additional, post-generation balancing option. They can store excess renewable energy for use when there is a shortfall. Hydro pumped reservoir storage is the main option so far and, as just indicated, there are also tidal barrage and lagoon pumped storage options, but there are many other storage possibilities, as the next chapter will explore.

Chapter summary

1. Grid systems mainly deal with variable supply and demand by ramping gas-fired plant output up and down and, in some cases, by using pumped hydro storage systems.
2. Adding renewables to the supply mix means that gas plant ramping will have to occur more often, involving a small extra cost penalty and a small reduction in the emissions avoided by using renewables.
3. With moderate levels of renewables on the grid, there will be little need for extra back-up: the necessary balancing capacity already exists in the rest of the system, although nuclear plants are less flexible and may not be able to help much.
4. With larger inputs from variable renewables, and fossil fuel use reduced, additional balancing systems will be needed, but, along with the increasing use of non-variable renewables, it may be possible for the use of fossil fuel back-up to be limited.

References

Abbess J 2015 *Renewable Gas* (Basingstoke: Palgrave)

Aris C 2014 *Wind Power Reassessed: A Review of the UK Wind Resource for Electricity Generation* (London: Adam Smith Institute/Scientific Alliance) www.adamsmith.org/wp-content/uploads/2014/10/Assessment7.pdf

Bade G 2015 PowerGen 2015: Why 'capacity will no longer be the coin of the realm' in the power sector *Utility Drive* (10 December 2015) www.utilitydive.com/news/powergen-2015-why-capacity-will-no-longer-be-the-coin-of-the-realm-in-th/410613/

Bird L, Cochran J and Wang X 2014 *Wind and Solar Energy Curtailment: Experience and Practices in the United States* (Golden, CO: National Renewable Energy Laboratory) www.nrel.gov/docs/fy14osti/60983.pdf

Boyle G (ed) 2007 *Renewable Electricity and the Grid* (London: Earthscan)

Cany C, Mansilla C, Da Costa P, Mathonnière G and Thomas J-B 2016 Nuclear power: a promising back-up option to promote renewable penetration in the French power system?

Renewable Energy in the Service of Mankind Vol II, Selected Topics from the World Renewable Energy Congress WREC 2014 ed A Sayigh (Berlin: Springer) pp 69–80

Dale L, Milborrow D, Slark R and Strbac G 2004 Total cost estimates for large-scale wind scenarios in UK *Energy Policy* **32** 1949–56

Douglas J 2015 *Heat Insight: Smart Systems and Heat—Decarbonising Heat for UK Homes* (Birmingham: Energy Technologies Institute) www.eti.co.uk/wp-content/uploads/2015/03/Smart-Systems-and-Heat-Decarbonising-Heat-for-UK-Homes-.pdf

EDF 2008 *UK Renewable Energy Strategy: Analysis of Consultation Responses* (submission to the UK government's renewable energy strategy consultation, prepared for the Department of Energy and Climate Change, log no. 00439e) www.berr.gov.uk/files/file50119.pdf

Elliott D 2015 Tidal power—still moving ahead *Advances in Energy Research* **vol 22** (New York: Nova Science)

ERP 2015 *Managing Flexibility whilst Decarbonising the GB Electricity System* (London: Energy Research Partnership) http://erpuk.org/wp-content/uploads/2015/08/ERP-Flex-Man-Full-Report.pdf

Fraunhofer 2014 *Electricity Production from Solar and Wind in Germany in 2013* (Freiberg: Fraunhofer Institute for Solar Energy Systems) www.ise.fraunhofer.de/en/downloads-englisch/pdf-files-englisch/news/electricity-production-from-solar-and-wind-in-germany-in-2013.pdf

Grubb M 1991 The integration of renewable electricity sources *Energy Policy* **19** 670–88

IEA 2011 *Harnessing Variable Renewables* (Paris: International Energy Agency) www.iea.org/publications/freepublications/publication/harnessing-variable-renewables.html

House of Lords (UK) 2008 The economics of renewable energy *House of Lords Select Committee on Economic Affairs Report (November)* www.publications.parliament.uk/pa/ld200708/ldselect/ldeconaf/195/19502.htm

Loughton M 2007 Variable renewables and the grid *Renewable Electricity and the Grid: the Challenge of Variability* ed G Boyle (London: Earthscan)

Martin A 2013 Flexitricity presentation *Global Energy Systems Conf. (Edinburgh)* http://global-energysystemsconference.com/wp-content/uploads/presentations/GES2013_Day2_Session2_Alistair_Martin.pdf

Mills A, Botterud A, Wu J, Zhou Z, Hodge B-M and Heaney M 2013 *Integrating Solar PV in Utility System Operations* (Argonne, IL: Argonne National Laboratory) https://emp.lbl.gov/sites/all/files/lbnl-6525e.pdf

NEA 2012 *Nuclear Energy and Renewables: System Effects in Low-Carbon Electricity Systems* (Paris: Nuclear Energy Agency, OECD) www.oecd-nea.org/press/2012/2012-08.html

NG 2016 How we balance the country's electricity transmission system *UK National Grid* www2.nationalgrid.com/UK/Our-company/Electricity/Balancing-the-network/

Ofgem 2014 *Electricity Capacity Assessment 2014* (London: UK Office of Gas and Electricity Markets) www.ofgem.gov.uk/ofgem-publications/88523/electricitycapacityassessment2014-fullreportfinalforpublication.pdf

Probert T 2011 Fast starts and flexibility: let the gas turbine battle commence *Power Eng. Int.* **19** 6

RAE 2014 *Wind Energy: Implications of Large Scale Deployment* (London: Royal Academy of Engineering) www.raeng.org.uk/windreport

Sansom R and Strbac G 2012 The impact of future heat demand pathways on the economics of low carbon heating systems *BIEE 9th Academic Conf. (Oxford, Sept 2012)* www.biee.org/downloads/the-impact-of-future-heat-demand-pathways-on-the-economics-of-low-carbon-heating-systems/

Sinden G 2005 *Renewable Electricity Generation and Electricity Storage* (Presentation, Cambridge University, Energy Storage seminar, June 2005) www.cambridgeenergy.com/archive/2005-06-22/CEF-Sinden.pdf

Wilson G, Rennie A and Hall P 2014 Great Britain's energy vectors and transmission level energy storage *Energy Procedia* **62** 619–28

Wilson G 2015 Tesla batteries might power your home but stored fuels will still run the country *The Conversation* 2 July 2015 https://theconversation.com/tesla-batteries-might-power-your-home-but-stored-fuels-will-still-run-the-country-42859

Yates N, Walkington I, Burrows R and Wolf J 2013 Appraising the extractable tidal energy resource of the UK's western coastal waters *Phil. Trans. R. Soc.* A **371** 20120181

IOP Publishing

Balancing Green Power

David Elliott

Chapter 3

The next challenge: energy storage

Energy can be stored in a variety of ways and at various scales for both short and long term grid balancing. Smaller-scale short-term options include batteries, larger, longer-term, bulk storage options include pumped hydro reservoirs. Energy storage is generally expensive, but provides a way to deal with surplus outputs from renewable energy generators, enabling the excess energy to be used later, when there is lull in renewable output and/or a rise in demand. Storing gas or heat is easier than storing electricity, so it may be best in some situations to convert electricity generated from renewable sources into heat, liquid air or synthetic gases, although there will be conversion losses. Solar or biomass generated heat can be stored directly for subsequent use in meeting heating needs at various scales. Larger heat storage systems are usually more cost effective, as is true for most forms of storage, but some domestic-scale storage options may be viable.

3.1 Storing energy

Energy storage is basically about time-shifting post-generation electricity or heat, collecting it whenever it is produced and delivering it later when it is needed. Some have portrayed storage as essential if renewables are to be viable. Thus in July 2015, the UK Energy and Climate Change Secretary, Amber Rudd, said 'as we all know, until we get storage right, renewables are unreliable' (Rudd 2015). That may be overstating the case. As we saw in the last chapter, there are already effective and widely used ways to balance variable renewables, and they can be extended as renewables expand, at relatively low costs. By contrast, storage is expensive, in part since, while existing power plants can be used for balancing with little extra cost, in most cases storage systems have to be built and, by their nature, these costly extra plants only deliver electricity for part of the time.

As argued in the previous chapter, it is much cheaper to store fuel before generation. However, in the case of renewables like wind and solar, that is not an option. These natural energy flows will of course still be there if they are not used, just like fuel stocks, but once they are converted into electricity, this has to be used, stored, or, if neither of those options is possible, wasted (Wilson *et al* 2014).

Storage can be expensive, in most cases costing much more than electricity generation, but making a direct comparison is hard, since it depends on how often the store delivers output and for how long. The retail costs of grid electricity can at present range from roughly 0.1–0.2 US dollars kWh^{-1}, sometimes less, sometimes more, depending on the sources and the country. For comparison, recent improvements in lithium ion batteries have brought their cost down to around 3–4000 US dollars for a 10 kWh rated storage system, like that developed by Tesla (Tesla 2015). If recharged, that could deliver 10 kWh regularly, for maybe 1000 cycles over its lifetime (before its performance degraded), so that the electricity from it would cost 0.3–0.4 US dollars kWh^{-1} (Naan 2015). At root, this large cost difference compared with generation costs is because the generally high capital cost of storage has to be recouped from, in most cases, relatively short, although hopefully multiple, periods of sometimes limited output. Table 3.1 illustrates this, and also indicates some other key characteristics of storage systems: some have limited storage capacity/energy storage densities.

For more details on storage systems and costs, see the US Energy Storage Association (ESA 2016), from which the table 3.1 data were obtained, the European Association for the Storage of Energy (EASE 2016) and the review from the UK Parliamentary Office for Science and Technology (POST 2015). IRENA also produced a useful report on storage in relation to renewables (IRENA 2015).

As these sources point out, storage can make sense in some situations, for example when there are large amounts of relatively cheap or valuable energy

Table 3.1. Electricity storage options—typical costs (in US dollars) and some key characteristics.

Pumped storage:	
very high capacity, low cost, but site specific	$100 kW^{-1}
Compressed air storage:	
high capacity, low cost, but site specific	<$100 kW^{-1}
Flow batteries:	
high capacity, but low energy density	$100–1000 kW^{-1}
Metal–air batteries:	
cheap, very high energy density, but not rechargeable	~$50 kW^{-1}
NaS, Li-ion and Ni–Cd batteries:	
expensive, but Li-ion now getting cheaper	~$1,000 kW^{-1}
Lead–acid batteries:	
cheaper, but bulky and limited deep cycling life	<$1,000 kW^{-1}
Flywheels:	
high power, but low energy density	>$1,000 kW^{-1}
Capacitors:	
high efficiency, cheap, but low energy density	>$100 kW^{-1}
High power capacitors:	
very expensive	<$10,000 kW^{-1}

available that would otherwise be wasted, as can be the case with the occasional surplus generation from some renewables. That is why pumped hydro storage, already used for system balancing, is also now being used for storing excess wind-derived electricity. The hydro reservoirs already exist, although not all are currently pump storage-enabled, and it is an option for them (JRC 2013).

In the case of variable renewables, storage like this can be economically viable, since it also provides a way to supply energy when their output is not available, and in some situations that may be a better option than finding interim fill-in replacement fossil sources, given that they will produce emissions. Certainly, as the proportion of renewables expands, storing green energy will become more attractive, since it avoids these emissions. It can also make system operations more flexible and avoids the use of increasingly costly fossil fuels in back-up plants.

The case for storage has certainly become stronger in recent years. In 2012, Dr Tim Fox, Head of Energy and Environment at the Institution of Mechanical Engineers, launching their report on storage, said: 'For too long we've been reliant on using expensive 'back-up' fossil-fuel plants to cope with the inherent intermittency of many renewables. Electricity storage is potentially cleaner and once fully developed is likely to be much cheaper'. The IMechE also lists some other positive benefits: storage units are more modular and flexible, with faster start up rates than some back-up plants, and can be used at various scales and locations (IMechE 2012).

With new technology emerging, the case for storage certainly seems to be improving, for batteries especially, driven by developments in the electric vehicle field (Nykvist and Nilsson 2015). Storage may therefore begin to challenge gas plants in their supply and balancing roles, peak demand standby plant especially. As a US observer put it: 'Peaker plants are reserve natural gas plants. On average they're active far less than 10% of the time. They sit idle, fueled, ready to come online to respond to peaking electricity demand. Even in this state, bringing a peaker online takes a few minutes. Peaker plants are expensive. They operate very little of the time, so their construction costs are amortized over few kWh; they require constant maintenance to be sure they're ready to go; and they're less efficient than combined cycle natural gas plants, burning roughly 1.5× as much fuel per kwh of electricity delivered, since the economics of investing in their efficiency hardly make sense when they run for so little of the time. The net result is that energy storage appears on the verge of undercutting peaker plants' (Naan 2015).

3.2 Battery storage

We will be looking at the relative merits of the various balancing options later on (storage and fossil back-up are not the only ones), but for now, we can simply observe that whether the cost of storage is seen as prohibitive or not in specific circumstances may often reflect issues of scale, utility and convenience. Certainly, storage generally makes sense if there is no other suitable source of energy available. We are happy to pay over the odds for energy from small, convenient batteries, a huge amount in kWh terms, to power torches or other small portable devices, since there is no alternative (clockwork wind-up springs apart). However, for most other purposes, at present, with pumped hydro being the main exception, the cost is usually seen as too high, while other balancing options are available. That assessment may be changing. The advent of lower cost lithium ion batteries, initially developed for electric vehicle use, and the publicity surrounding some specific examples, like Tesla's Powerwall, have put storage more centrally on the agenda.

As we have seen above, even the best modern batteries are not cheap. Tesla's 7 kWh domestic-scale version was launched in 2015 at a retail price of 3000 US dollars, excluding installation costs (Tesla 2015). Consumers wishing to use it to store electricity from a rooftop PV solar array would also have in install an inverter unit. Assuming there was sufficient daytime solar input, that system might allow them to run a 1 kW rated heater overnight. That may seem a small gain, but batteries like this can also reduce the PV capacity needed by consumers, or the import top-ups they require, in order to meet demand peaks at other times. So there may be other gains. Moreover, the technology is developing rapidly. The economics of domestic scale PV solar plus batteries may improve, making home storage a more popular option, perhaps as a way to be free of reliance on grid supplies from utility companies. They could perhaps even go 'off grid'.

Whether that would be an attractive option depends on how consumers actually use their domestic self-generation systems. At present many domestic PV users enjoy feed-in tariff or net metering arrangements, some of which allow them to earn income by selling surplus electricity to the grid. That can offset the cost of the PV array. If consumers store their excess for their own (later) use, when they have a shortfall, they may reduce their imports, but lose this lucrative export option. So it may not make sense for them, unless they want to go entirely off grid, balancing their own energy use with no top-ups from, or exports to, the grid. Some, of course, may want to do that, for example to be free of utility entanglements, and some say 'grid defection' will become common, especially in the USA (RMI 2014).

However, most consumers in the developed world are on the grid and grid defection on a large scale seems unlikely to spread, except in remote areas, since in many places it might be hard (and expensive) to balance energy use over the year just from PV solar and battery back-up. So grid links are likely to remain the norm, even with storage in place. Indeed, in system terms, there may be some value to having a network of domestic-scale storage units linked via the grid. That could act as a distributed energy store, if consumers were willing to let energy flow in and out when needed. It has also been suggested that the batteries in electric vehicles could be used in this way, in so-called 'vehicle to grid' mode (V2G). The use of home and car batteries would help with overall grid balancing, reducing the need for curtailment of surplus output, and providing energy for the grid when there is a shortfall and/or when there are demand peaks.

There are some uncertainties about these consumer-to-grid electricity exchange approaches. Perhaps the most obvious is that consumers may not be happy to have their home and/or car batteries drained when there is a grid shortfall, even if they are given a special tariff to compensate them for offering this service. There are also more indirect system issues. One system-level attraction of domestic stores, used just by the consumer to store excess electricity they have self-generated, is that this will reduce the amount of electricity that is transmitted on the local grid. So grid management and upgrade costs would be reduced. However, if consumers open their stores to interactive grid use, more electricity will flow (in and out) and the local grid may have to be strengthened to cope. With V2G, most of the charging power for the vehicle batteries would be coming in from the grid, overnight, while the car was parked, not from rooftop PV, unless the householder had a large separate battery store charged during the day. In the absence of that, the local distribution grid may have to be strengthened to cope with electric vehicle (EV) battery charging and also any battery-to-grid outflow. A UK study suggested that peak electricity demand could rise 20% with large-scale EV charging (DECC 2008). However, not everyone would necessarily be charging at the same time, and it has also been argued that the EV peak could be softened by sensible management (Heron 2015). Even so, clearly electric vehicle use and V2G open up many issues (see box 3.1).

In addition to the practical problems discussed in box 3.1 for the case of V2G, some interesting potential market conflicts can also emerge with grid-linked domestic storage. In situations where off-peak electricity is sold cheaply on the grid, domestic scale storage can in theory also be used to store it, and, if the consumer has a suitable net metering tariff, they could in theory sell it back to the grid at a profit. That has obviously worried utilities, some of whom have denied contracts to consumers with stores (Poole 2013).

From the consumer's point of view, the store-and-resell practice makes economic sense, and indeed it can also make sense from the overall energy system point of view: energy flows are better balanced. However, from the utilities point of view, it undermines their profits. This issue has come to a head in Spain, where in 2015 a new tax was imposed on the use of PV solar with storage. Ostensibly, it was because PV consumers were making use of the grid without charge, but opponents have seen it as an attempt to tax the use and storage of sunlight (Pentland 2015).

The role of the large energy utilities will certainly be changed if local storage and local distributed generation spreads. The growth of 'prosumer' self-generation and local energy co-operatives in Germany has already had a major impact on utilities there, as we will see in chapter 4. Storage would add to that. Utilities may have to adjust by becoming energy service companies, supporting local generation and storage, rather than just being centralised energy suppliers (Corneli and Kihm 2015).

The debate over the merits of local-scale generation and domestic-scale batteries continues, driven in part by the advent of new technologies. There are many new ideas emerging for new battery systems at all scales and for a wide range of uses. While new types of more efficient lithium–air batteries are emerging from the research labs (Hoster 2015), some look to aluminium-ion batteries as a less toxic alternative (Lin *et al* 2015). Clearly, health and safety will be important issues, especially if the focus is on home use. In that context, one interesting development is a non-flammable flow battery that uses (blue) cake dye-type chemicals (Radford 2015). In parallel, there are developments in the supercapacitor field that may yet outflank batteries in some applications, since they can have higher energy densities than lithium ion batteries and potentially have fewer fire/toxicity hazards (IDTechEx 2014).

Box 3.1. Vehicle to grid balancing.

The **vehicle to grid** (V2G) concept offers a way to balance variable renewables and also demand peaks by using the batteries of electric vehicles linked to the grid when parked at home (or perhaps elsewhere) to store excess electricity during low demand periods, ready to be exported when demand is high and renewables low. It sounds a clever idea, but in addition to economic issues (e.g. the extra costs of the home-based electricity uploading system), it opens up some interesting logistical issues.

In the UK, peak energy demand coincides roughly with coming-home-from-work time (6–7 pm for many people), so many cars will not be 'nested' and linked-in at home. There would be a steady stream of arrivals from work, but most would have low charge, having just been driven. So when they arrive home they will be energy-hungry and not available for V2G power-to-grid uploading. Indeed, some see the arrival of millions of EVs in the evening as being a major issue, adding to evening demand peak. Prices for this period could be set high as a disincentive. There have already been trials of an incentive system like this in the USA (Hull 2015).

However, some US work has suggested that the demand peak issue may not be as big a problem as expected (Overton 2013). Certainly, those cars that are nested at home all day could be trickle charged during daylight hours from domestic PV, in summer especially, and then be (partly) discharged if necessary into the grid at peak demand times. The only problem is that drivers may not be happy if they decide to go out later in the evening (or next morning) and find the car almost totally discharged. There would have to be negotiated/contracted-for limits, and opt-outs. This is also important, since it has been claimed that regular extra deep cycling of batteries will age them rapidly, thus costing car owners more for more regular replacements, given that the batteries are the most expensive element in EVs. A wear and tear element has to be included. Indeed, it might be that it would be best for V2G-enabled cars to have more robust outsized batteries, designed for this function. That would add to the cost and would only be worth it for car owners if the cash value of V2G exports is high. That income, and the scale of use of the V2G option, would also be variable around the year, making investment assessments harder, whatever size or type of battery is used. That said, variable income from feed-in tariffs has not put off home-owners from investing in PV solar, so the same may be true for V2G and, in theory, despite the occasional electricity drain inconvenience, V2G income should offset the cost of running an EV and the cost of PV.

Nevertheless, looking at the energy system as a whole, some say the V2G idea is at best marginal and opportunistic: if you want to store energy, have proper dedicated stores. While it may seem clever to use all those batteries, which are mostly just sitting in parked cars, unused perhaps 80% of the time, there is also the fact that energy conversion losses using standard EVs will be high. Charging a battery system from the grid is around 70–80% efficient. Returning that energy from the battery to the grid, which will require converting the DC back to AC with efficiencies of about 90%, yields 63–72% overall system efficiency.

Other larger-scale bulk storage options may be better, e.g. pumped hydro reservoirs, except that, if the storage option is to expand, new stores will have to be built and connections made to them: the beauty of V2G is that the batteries already exist (or will soon) and V2G makes use of the same grid links already used by homes (though they may have to be reinforced). There is also the point that their output does not have to be used just for major peak matching: there will also often be small input and output perturbations on the grid, for example due to variable renewable inputs, that need to be balanced, and this can be done without decreasing the life of the V2G battery. Dealing with small ups and downs is a gentler process. So there may be a less strenuous role for V2G (and grids), operating outside of peaks.

To improve the (time) spread availability for battery charging in vehicles, it is conceivable that charging could be done in the daytime at the workplace, e.g. there are some large PV arrays on solar shades at car parks in the USA. Overall energy matching is also easier if, rather than relying just on PV for charging, wind-derived grid electricity is used. That can be available at most times of the day and night, in varying amounts, from wind farms, though some of the logistic issues still apply. Drivers will not want their batteries drained without warning, and charging and exporting may still strain the grid.

Whatever the charging source, there may also be grid-straining synergistic timing problems and conflicts with other energy uses. The UK government's current plan is for most houses to be heated electrically rather than by gas. However, if vehicle charging is required in the same dwellings as those containing electric heat-pumps, used for evening heating, then as the Energy Technologies Institute has noted, 'not only will the local sub-system be very likely to be overloaded, but the individual connection to the building may need to be replaced'. Moreover, nationally, with two systems requiring electricity, the evening peak demand will be raised even more. We will be following up that issue later.

Will V2G happen? The viability of the V2G idea relies on wide uptake of electric vehicles, which may take time, or not occur on a significant scale. However, some analysts have been very optimistic. For example, Zpryme's 'Smart grid insights' projected that by 2020 there could be over one million V2G-enabled vehicles on the road in the USA, with an estimated market value of 26.6 billion US dollars, and an additional 6.7 billion infrastructure market (Siegel 2010).

There are certainly some interesting projects and programmes going ahead in the USA and EU. However, it may take time for V2G to spread. In 2011, the UK National Grid said that it did not consider 'vehicle to grid (V2G) services, as economically viable in the near term due to additional costs that be incurred to make them export capable' (NG 2011). That assessment may change. Being optimistic, the Centre for Alternative Technology's 2030 *Zero Carbon Britain* scenario had 25 GWh annually coming from V2G EV batteries. Useful, but not huge (ZCB 2014). Moreover, even the high 'level 3' setting in the UK government's 2050 Energy Pathways model only has 5.3 GW of V2G by 2050 (DECC 2012). That assessment may also change, but for the moment the prospects for V2G in the UK look marginal.

It is certainly the case that battery and advanced storage system technology is a rapidly expanding field, with new applications opening up widely, including batteries for grid support and balancing, as well as for consumer use (Insight-e 2014). However, while home-scale applications may be important, in energy supply system terms, larger utility-scale storage may be more economical and efficient than domestic units, and in general that is usually the case with all storage systems.

3.3 Larger-scale storage

Scale is a key issue, in part because large systems can service the needs of large numbers of users, whose individual energy use patterns may differ. Meeting collective needs with large bulk stores, rather than each household having its own small store, averages these variations out: there are technical and operational economies of scale, including management and maintenance benefits.

To some extent, the larger the store, the better, although that depends on locational factors. Clearly, as already noted, large hydro reservoir-based pumped storage plants can only be in geographically defined areas. The same would apply to the use of tidal barrages or lagoons for pumped storage. It is the same for another bulk energy storage idea that is now being developed, compressed air storage. It is possible to store energy as compressed air on a small scale in engineered cylinders, but it is more economical to do this on a larger scale, and to use existing underground caverns, which can be suitably sealed to maintain pressure. Air can be pumped into them under high pressure using electricity from renewable sources (e.g. nearby wind turbines) and then let out through an air turbine to generate electricity when it is needed. In one version of this compressed air energy storage (CAES) approach, the compressed air is used to supercharge a conventional natural gas-fired turbine, boosting its energy output (Gaelectric 2015).

It should be noted that hydrogen gas can also be stored in underground caverns or aquifers. Natural gas is already stored in large quantities in this way in some locations (Natgas 2013). There are obvious safety issues (as was demonstrated by the major leak of methane from a large underground store in California in 2015/16), requiring attention to proper containment, but, as we shall see later, some say that underground hydrogen storage could be one of the cheapest bulk storage options. Moreover, some see hydrogen, produced from renewable sources, as being a key new energy vector, with large stores being linked to hydrogen pipelines, or tankers/trucks being used for distribution, depending on end-use requirements (Andrews and Shabni 2012).

However, suitable underground caverns for storing air or hydrogen only exist in a limited number of places, and, like pumped hydro, this is large-scale approach, requiring long-distance electricity grid transmission (or pipelines, in the case of hydrogen) to users. For example, a very ambitious 8 billion US dollar project has been proposed in the USA, which involves a 2.1 GW wind farm in Wyoming sending electricity by a high-voltage direct current (HVDC) power grid 525 miles to an underground salt cavern compressed air storage facility in Utah, and then, after conversion back to electricity, 490 miles on to Los Angeles (PennEnergy 2014, Gruver and Brown 2014).

There may be attractions in having somewhat smaller storage systems nearer to end-use loads or to renewable supply systems, so as to avoid long-distance electricity transmission and improve system flexibility. In terms of siting near generators, one novel idea that has emerged in the context of offshore wind projects is to use giant undersea inflatable bags tethered nearby to store compressed air generated using electricity from the turbines. The air would then be used to drive an air turbine when electricity was needed. Round-trip efficiencies of 90% have been claimed for this concept (Garvey 2011).

Some medium-scale battery projects are going ahead, including a 5 MW/5 MWh project in Germany with lithium–manganese–oxide cells, and a 2 MW system in the

UK, but other storage technologies are emerging that may also be relevant for multi-megawatt utility-scale storage. For example, it is possible to use excess renewables-derived electricity to liquefy air. This can be stored for later use for electricity generation, with the liquid air being warmed to produce a high-pressure airflow to drive a turbine. A system like this can be located anywhere, although there are obvious advantages in siting it near a source of heat for the revaporisation process; waste heat from a conventional power plant might be an example (Highview 2015).

Other medium-scale storage options include the new generation of flow batteries, in which energy is stored in separated chemical electrolytes and released when they are re-combined, in a reversible process (VRB 2012). On the horizon there are non-toxic organic variants (Harvard 2014). Some larger versions of more conventional batteries are also being developed, such as sodium sulphur batteries. And Tesla is offering a utility-scale multiple-unit version of its Li ion battery system, with other suppliers also developing rival variants, the idea being to have large grid-linked battery storage parks.

Scale and locational issues are of course not the only factors. To avoid physically very large space-consuming stores, high energy densities are helpful, but what also matters, for economically viable storage, is how long it takes to charge a store, the length of time that energy can be stored in the system without undue losses, and how quickly it can be released. There are trade-offs between capacity and speed of discharge (figure 3.1). Some storage technologies also have different charging and discharging capabilities. For example, an IEA study noted that 'CAES systems generally have twice the up-ramping capability compared to down-ramping manoeuvrability i.e. they can produce electricity faster than they can store it.

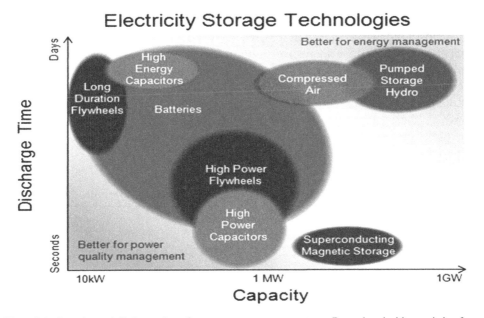

Figure 3.1. Capacity and discharge times for some energy storage systems. Reproduced with permission from EIA (2011).

By contrast, electrochemical storage in batteries has comparable values for production as for charging' (IEA 2005).

Some of the larger systems, like pumped hydro and CAES, can store energy for long times, and depending on their scale, can also continue to deliver energy for long times. The UK's Dinorwig pumped hydro scheme, if fully charged, can supply electricity at full power for 5 h, and respond very quickly to shortfalls. The smaller systems are usually easy to load up and discharge, but are often more expensive per kWh, and are not so useful for long-term storage. They may be best used for fast and regular short-term storage, for example for dealing with daily demand peaks, rather than larger and maybe longer unexpected shortfalls.

However, perhaps the key factor is the overall energy round trip efficiency of the process, from input to output. Systems that involve multiple energy conversions may be problematic. At each stage there may be significant losses. See table 3.2 for estimates of energy conversion efficiency for some common storage systems of varying size. Conversion losses are certainly an issue that one of the more exciting new energy storage options faces, storage as hydrogen gas. As table 3.2 shows, if it is then used in a fuel cell to generate electricity, the overall efficiency can be low.

Nevertheless, the hydrogen option is worth looking at in more detail, since it offers some interesting possibilities for bulk storage using surplus electricity from variable renewables. Hydrogen can be produced by the electrolysis of water using electricity from a renewable source, and then stored as a gas or cryogenic liquid, or perhaps in chemi-absorbed form, while underground caverns are a storage option. It can then be used to make electricity again, by firing a gas turbine or feeding into a fuel cell. Electrolysis is about 70% efficient, with the losses emerging as heat, but some of that can be captured and used, pushing up the overall energy efficiency of the process. Cryogenic storage has losses: it takes energy to liquefy gases, and some is also lost/used when they are released as vapour. Fuels cells can be 40–50% efficient, or more with heat recovery, and combined cycle gas turbines might be similar. If you add up all these losses, the overall process may be well under 40% efficient. That may not matter too much if the original energy source is in effect free, assuming it is surplus renewable electricity that would otherwise be wasted. However, there may actually be other uses for the hydrogen than using it to generate electricity. For example, it can be converted into other fuels, which may

Table 3.2. Storage technology: key characteristics. Reproduced with permission from IEA (2005).

Storage technology	Typical round trip efficiency	Typical capacity
Pumped hydro	~80%	>100 MW–1000 MW
Compressed air	~75%	>50 MW–100 MW
Flywheel	~90%	>1 kW–50 kW
Conventional batteries	~50–90%	>1 kW–>10 MW
Flow battery	~70%	~15 MW
Hydrogen fuel cell	~40%	>50 kW–>1 MW

have higher value. The hydrogen can be reacted chemically with carbon dioxide gas captured from the air, or more likely (since the proportion of CO_2 is much higher) from power station exhausts, to make methane gas, methanol, or other synthetic fuels, for example for use in vehicles.

This so-called power-to-gas concept, in its various forms, has obvious attractions. It offers a way to turn a problem (surplus electricity) into a solution (new fuels) for a variety of possible uses, including balancing the grid, heating, and powering vehicles. So, as figure 3.2 illustrates, it offers a range of flexible end-use options.

As figure 3.2 shows, hydrogen can not only be used directly for electric power production, it can also be injected into the gas grid (admixed with fossil gas), as can synthetic methane, of course, reducing the need for electricity transmission. This is already being done in Germany, and this role could expand, as could its role in direct balancing via electricity generation. A German federal government study noted that, as renewables expand in Germany, 'situations could occur in some regions that would prevent the distribution grids from coping with such high levels of electricity. In these regions, power-to-gas technologies could start to make sense in just a few years' time, offering an easily regulated load that can provide potential capacity on the backup power market and help take pressure off the grid' (UBA 2013).

While the power-to-gas idea has attractions as a balancing option, enabling surplus electricity to be used to meet lulls, and that may grow, for the moment the main commercial focus is on power-to-gas (P2G) and power-to-liquid (P2L) conversion for vehicle fuel production, since that has high value.

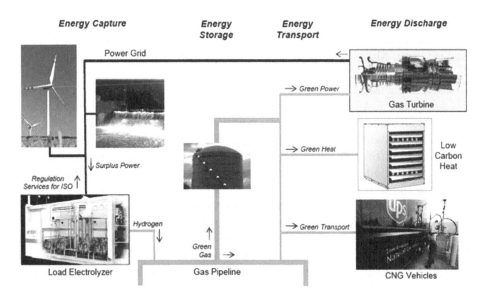

Figure 3.2. Power-to-gas conversion and end-use options. Reproduced with permission from Hydrogenics (2012).

As can be seen, it may thus be the overall system economics, coupled with the wider energy market, that decide what approach is taken. We will be coming back to that in a later chapter, when we look at how the various energy storage and grid balancing options may interact within an integrated energy system. For the moment, however, in terms of energy storage options, it is important to note that there can be more to it than storing electricity or gas, or pumping water to create a head. Storing heat is arguably a much easier option.

3.4 Heat storage

Hot water can hold about 3.5 times as much energy by volume as natural gas at atmospheric pressure and temperature. As with other forms of storage, heat storage is usually best done on the larger scale, balancing-out individual house-heat use variations, with the added point being that, in large heat stores, the heat losses through the surface area are lower than those from smaller stores, since the ratio of area to volume falls with increasing size. Large community heat stores, fed partly by solar heat, are now quite widespread in Denmark and northern Europe, some of them allowing summer heat to be used for winter warming. Certainly storing heat for longer periods has obvious attractions where sunshine is not guaranteed and the climate can be cold, but there are limits to the size of the hot water store that an individual house can accommodate. That, as well as the economics, also makes larger community-scaled stores, which can be located nearby, attractive.

Although large-scale systems like this seem to make most sense, there is nevertheless some interest in smaller-scale heat storage for individual houses or other buildings. Rooftop solar water heaters for space, or more usually water, heating are widely used around the world, most obviously in sunny climate zones, but also sometimes in areas less blessed with warm sun: in far northern (or far southern) climates, the value of any solar energy that can be obtained is higher and the heating season is longer. Most of these systems feed the heat they collect into a domestic hot water store, linked to the hot water system, for daily use.

Using direct solar heat is not the only option. Photovoltaic solar cells do not need warm sun to run, just strong daylight, and in some locations they can produce more electricity than consumers need at some points in the year. So, as noted earlier, some consumers feed this to domestic-scale batteries. However, there is another option. The surplus electricity from consumers' rooftop PV arrays can be fed to immersion heaters in their hot water boilers. Whether this hot water storage option makes economic sense is unclear. If the consumer has signed up to a feed-in tariff or net metering system, it will probably make more sense to export any excess PV electricity rather than storing it as heat, or, for that matter, directly in batteries, although the use of larger heat or electricity stores might alter that assessment. For example, there are some interesting community-scale PV solar experiments with large shared battery storage systems (ERIC 2016). However, for heat, direct solar thermal panels are likely to be more economical, depending on location, and grouped solar schemes, as described above, with large shared community heat stores, would seem to be the best solar heat option in the urban, suburban or town context.

Local solar heating and storage is not the only way in which heat can play a role in balancing energy availability and use over time. Heat from a range of sources can also be transmitted over relatively long distances. There are energy losses, but there is a 65 km large diameter overland heat link pipe between a biomass waste-burning plant and the city of Prague. Urban district heating networks that are many kilometers long, with smaller lagged pipes buried underground, are common in parts of the EU, with many of them being fed by CHP plants, and some of them using biomass or wastes as a fuel. The district heating pipes act as a heat store, and the systems may also have large linked heat stores for balancing heat demand. The overall energy conversion efficiency can be very high, 80% or more, since the CHP plants make use of heat that would otherwise be wasted. So, even if they use fossil fuels, they can be seen as relatively low carbon plants. If they use sustainable biomass as a fuel, then the emissions fall even more, with, for example, sewage biogas and landfill gas being some of the cheapest energy sources around. That makes this approach to energy supply increasingly attractive (see box 3.2).

Of more direct relevance to our concerns here is the fact that systems like this, with hot water storage capacity included, can also offer an interesting grid balancing option. CHP plants can vary the ratio of heat to power output. If electricity demand is low and there is too much coming into the grid from renewables, the CHP plant can produce more heat and less power, and, if heat demand is also low, store it until heat demand is high. So CHP plants can help limit the need for the curtailment of excess renewable output. They can also help if there is a renewable shortfall. If the renewable supply is low, the CHP plant can produce more power and less heat, and if heat demand is also high, it may be able to supply it from the store, if that has been charged. The heat stores can also be fed with solar heat or heat produced by immersion heaters powered with surplus renewable electricity, for example from wind farms (JRC 2012).

Hot water storage is not the only heat storage option. Heat can also be stored in other media, including crushed rocks and gravel. The pumped heat electricity storage system developed by Isentropic in the UK involves pumping heat from a cold, gravel-filled container into a hot one. Reversing this flow, when output is needed, drives a heat engine, which generates electricity (Isentropic 2016). Molten salt heat stores are also being used, with large concentrated solar power (CSP) plants in desert areas, allowing for continued electricity supply overnight. Some of the daytime solar heat is stored to produce steam for running the turbines at night. Other storage media are also being investigated for use with CSP, including sand (Williams 2014), of which there is clearly plenty to hand in desert areas, and concrete matrix materials (Barragán and Schmitz 2015). In the case of wind turbines, Siemens was reported to be looking at rock heat storage (Smith 2014). A perhaps more exotic idea is the generation and storage of heat produced directly from wind turbines via mechanical heat churns or eddy current electromagnetic induction heating (Okazaki *et al* 2015).

The use of on-site heat storage for wind turbines and to allow CSP plants to run 24/7 is clearly a direct solution to intermittency issues and, as we have seen, heat storage can also help CHP plants to balance variable renewable inputs to the power grid, while feeding heat to users over extended networks. So heat storage and transmission may play a growing role in energy systems. Although it may be easier to supply houses with electricity or gas, heat is usually the largest element of their energy demand and, in some urban and suburban locations, supplying it directly over local networks, with large buffer stores balancing-out variations, may be a better option.

However, for long-distance energy supply, clearly gas and electricity transmission win out. Gas transmission may actually have the edge, in that losses are low and, with the pipes underground, it is less invasive, once installed, than overhead power grids. Moreover, gas is easy to store, unlike electricity. However, long-distance electricity transmission also offers some interesting grid balancing possibilities, and the next chapter will explore them.

Box 3.2. Home heating and CHP/district heating in the UK.

Domestic heating accounts for around 40% of UK energy use, but so far this has mostly been met with home-based appliances, such as gas-fired central heating boilers or electric or gas fires. More recently, some interest has also been shown in domestic-scale electric heat pumps, since they can be a more efficient way of using electricity. Heat pumps can use electricity to upgrade ambient heat, with a coefficient of performance (COP) of maybe three, meaning that the energy content of the heat output is three times that of the electrical input. However, community-scaled CHP plants, using otherwise wasted heat, have COP equivalents of up to nine or more, depending on the network temperature required (Lowe 2011).

Installing district heating networks to distribute this heat is disruptive and they only make sense in urban and some suburban areas, but there are clear advantages with larger community-wide district heating systems, as used widely across much of northern Europe (JRC 2012). In a 2012 report the Royal Academy of Engineering said that 'larger district systems, incorporating a CHP facility and providing heating are significantly more efficient than domestic level installations. Central systems may be more efficient and are likely to offer much greater energy storage than do systems designed for individual household' (RAE 2012). UK interest in CHP/district heating has now expanded, with the DECC backing heat networks. It said that 'some models show technical potential to supply as much as 43% of heat demand for buildings through networks by 2050' (DECC 2015).

It should also be mentioned that CHP plants, as well as other heat and power sources, can also be used to power cooling systems, for example using absorption chillers, and that district cooling networks have attractions in urban environments. Indeed, as climate change impacts further, cooling will become a key issue in many parts of the word. To the extent that grid electricity production is still dominated by fossil fuel-fired generation, running air conditioner units on mains electricity will just make the overall situation worse. PV solar-derived electricity could be used to run them, for example in daytime occupancy office buildings, but, in the domestic context, the need for air-conditioning may be mostly after the Sun has gone down, implying the need for battery storage. District cooling systems would perhaps be a better option, possibly linked to cool stores: it is just as efficient to store cold as it is to store heat. As noted above, liquid air is being explored as a storage medium for power generation systems, and there are many applications for new types of cooling technology in many other sectors (Birmingham Energy Institute 2015).

3.5 The way ahead

Energy storage is in the midst of a boom phase at present, with many new ideas emerging and older concepts being improved. Developments in materials science are one of the drivers, for example for batteries for electric vehicles, and also for use with domestic PV solar arrays. At the larger scale, in addition to compressed air storage in caverns, some quite exotic ideas have emerged, like the use of flywheels for utility scale energy storage: a 20 MW flywheel system is being built in Ireland at Rhode, County Offaly (DJEI 2015). There is also the 'gravity store' concept, with a large mass being lifted hydraulically in a large water-filled piston system in a vertical silo in the ground, and electricity being generated via turbines running on water squeezed out when the mass falls (Gravity Power 2014).

However, conventional reservoir-based pumped-hydro remains the favoured option at the utility scale, and there are interesting examples of direct use with renewables, like the reservoir system installed on the Canary Island of El Hierro to store the energy generated by five wind turbines (Frayer 2014). This involves conventional electricity generation by wind turbines and pumped hydro storage ready for electricity generation when needed.

The development of these and other technologies, including heat and hydrogen storage systems, is likely to open up new concepts for energy supply and use, and in particular energy transmission. Storage offers a way to avoid having to transmit electricity instantly to users, as does the use of gas and (to a lesser extent) heat as energy vectors. Nevertheless, conventional electricity grids are likely to remain the mainstay of energy systems for some while, and as the next chapter indicates, they offer a range of balancing options.

Chapter summary

1. Energy storage is expensive, but can help to deal with variable renewables at various scales, if the value of the energy saved is sufficiently high.
2. A range of electricity and heat storage options is available, which can store energy over a range of times and at various scales.
3. In general, there are economies of scale, so that community- or utility-scaled storage is often the most efficient option. But there can also be value in domestic-scale storage.
4. Gas and heat are easier to store than electricity and the demand for heat is usually larger and more variable (daily and seasonally) than that for electricity.

References

Andrews J and Shabini B 2012 Where does hydrogen fit in a sustainable energy economy? *Procedia Eng.* **49** 15–25

Barragán J and Schmitz M 2015 Heatcrete or molten salt? *CSP Today* (13 February 2015) http://social.csptoday.com/technology/heatcrete-or-molten-salt?

Birmingham Energy Institute 2015 *Doing Cold Smarter* (Birmingham: University of Birmingham) www.birmingham.ac.uk/Documents/college-eps/energy/policy/Doing-Cold-Smarter-Report.pdf

Corneli S and Kihm S 2015 Electric industry structure and regulatory responses in a high distributed energy resources future *US Lawrence Berkeley National Labs 'Future Electric Utility Regulation' Study Programme* report no. 1 https://emp.lbl.gov/future-electric-utility-regulation-series

DECC 2008 The impact of changing energy use patterns in buildings on peak electricity demand in the UK *Building Research Establishment Report* (London: Department of Energy and Climate Change) http://www.gov.uk/government/uploads/system/uploads/attachment_data/file/48191/3150-final-report-changing-energy-use.pdf

DECC 2012 *2050 Pathways Calculator* (London: Department of Energy and Climate Change) http://webarchive.nationalarchives.gov.uk/20121217150421/http://decc.gov.uk/en/content/cms/tackling/2050/2050.aspx

DECC 2015 *Delivering UK Energy Investment: Networks 2014* (London: Department of Energy and Climate Change) www.gov.uk/government/publications/delivering-uk-energy-investment-networks-2014

DJEI 2015 *First Hybrid-Flywheel Energy Storage Plant in Europe announced in Midlands* (Dublin: Irish Department of Jobs, Enterprise and Innovation) www.djei.ie/en/News-And-Events/Department-News/2015/March/First-Hybrid-Flywheel-Energy-Storage-Plant-in-Europe-announced-in-Midlands-.html

EASE 2016 *European Association for the Storage of Energy* www.ease-storage.eu/technologies.html

EIA 2011 Chart from US Energy Information Administration website, based on Energy Storage Association data www.eia.gov/todayinenergy/detail.cfm?id=4310

ERIC 2016 Energy Resource for Integrated Communities project (Oxford, UK) www.meetmaslow.com/wp-content/uploads/2015/05/MASLOW-60Kwh-Community-Case-Study-OXFORDy.pdf

ESA 2016 US Energy Storage Association http://energystorage.org/energy-storage/energy-storage-technologies

Frayer L 2014 Tiny Spanish island nears its goal: 100 per cent renewable energy *NPR* (28 September 2014) www.npr.org/sections/parallels/2014/09/17/349223674/tiny-spanish-island-nears-its-goal-100-percent-renewable-energy

Gaelectric 2015 CAES system www.gaelectric.ie/index.php/energy-storage/

Garvey S 2011 The dynamics of integrated compressed air renewable energy systems *Renew. Energy* **39** 271–92

Gravity Power 2014 *Gravity Power* www.gravitypower.net/technology-gravity-power-energy-storage/

Gruver M and Brown M 2014 Renewable energy plan hinges on huge Utah caverns Associated Press (24 September 2014) http://bigstory.ap.org/article/3084cb4c459f4ffd9b666f5d5d2e44e3/wind-energy-proposal-would-light-los-angeles-homes

Harvard 2014 Organic mega flow battery promises breakthrough for renewable energy *Press Release* (Cambridge, MA: Harvard University) www.seas.harvard.edu/news/2014/01/organic-mega-flow-battery-promises-breakthrough-for-renewable-energy

Heron D 2015 Evaluating the full transportation and energy life-cycle *The Long Tail Pipe* (16 April 2015) http://longtailpipe.com/2015/04/16/fast-charging-electric-cars-wont-swamp-electricity-grid-if-done-intelligently

Highview 2015 *Highview Cryogenic Energy Storage* www.highview-power.com

Hoster H 2105 Lithium–air: a battery breakthrough explained *The Conversation* (2 November 2015) https://theconversation.com/lithium-air-a-battery-breakthrough-explained-50027

Hull D 2015 Vehicle to grid energy storage experiment underway in California *Renewable Energy World* (5 August 2015) www.renewableenergyworld.com/articles/2015/08/vehicle-to-grid-energy-storage-experiment-underway-in-california.html

Hydrogenics 2012 Power to gas chart *Presentation by Hydrogenics CEO* www.hydrogen.energy.gov/pdfs/htac_may2012_hydrogenics.pdf

IDTechEx 2014 Supercapacitors can destroy the lithium–ion battery market *IDTechEx* www.idtechex.com/research/articles/supercapacitors-can-destroy-the-lithium-ion-battery-market-00006649.asp

IEA 2005 *Variability of Wind Power and other Renewables: Management Options and Strategies* (Paris: International Energy Agency) www.uwig.org/iea_report_on_variability.pdf

IMechE 2012 Policy statement on energy storage (Institution of Mechanical Engineers) www.imeche.org/knowledge/policy/energy/policy/ElectricityStoragePolicyStatement

Insight-e 2014 How can batteries support the EU electricity network? European Energy Report No. 1, INGIGHT_E, KTH Stockholm www.insightenergy.org/ckeditor_assets/attachments/48/pr1.pdf

IRENA 2015 *Renewables and Electricity Storage* (Abu Dhabi: International Renewable Energy Agency) www.irena.org/DocumentDownloads/Publications/IRENA_REmap_Electricity_Storage_2015.pdf

Isentropic 2016 *Isentropic* www.isentropic.co.uk

JRC 2012 *District Heating and Cooling* (Petten: European Commission Joint Research Centre) http://setis.ec.europa.eu/system/files/JRCDistrictheatingandcooling.pdf

JRC 2013 *Assessment of the European potential for PHS* (Petten: European Commission Joint Research Centre) http://setis.ec.europa.eu/newsroom-items-folder/jrc-report-european-potential-pumped-hydropower-energy-storage

Lin *et al* 2015 An ultrafast rechargeable aluminium-ion battery *Nature* **520** 324–8

Lowe R 2011 Combined heat and power considered as a virtual steam cycle heat pump *Energy Policy* **39** 5528–34

Naan R 2015 Why energy storage is about to get big—and cheap *Ramez Naan Energy Blog* (14 April 2015) http://rameznaam.com/2015/04/14/energy-storage-about-to-get-big-and-cheap/

Natgas 2013 Natural gas storage in the USA *Natgas* http://naturalgas.org/naturalgas/storage/

NG 2011 *Operating the Electricity Transmission Networks in 2020—Update June 2011* (London: UK National Grid) www.invictacapital.co.uk/electronic_mailshots/Store/NG_Operatingin2020_finalversion0806_final.pdf

Nykvist B and Nilsson M 2015 Rapidly falling costs of battery packs for electric vehicles *Nat. Clim. Change* **5** 329–32

Okazaki T, Shira Y and Nakamura T 2015 Concept study of wind power utilizing direct thermal energy conversion and thermal energy storage *Renew. Energy* **83** 332–8

Overton T 2013 Impact of electric vehicle charging on grid may be far less than feared *Power* (24 October 2013) www.powermag.com/impact-of-electric-vehicle-charging-on-grid-may-be-far-less-than-feared/

PennEnergy 2014 $8B renewable energy initiative proposed for Los Angeles *PennEnergy* (23 September 2014) www.pennenergy.com/articles/pennenergy/2014/09/8b-renewable-energy-initiative-proposed-for-los-angeles.html

Pentland W 2015 Energy storage is the real target of Spain's new tax on the Sun *Forbes* (18 June 2015) www.forbes.com/sites/williampentland/2015/06/18/energy-storage-is-the-real-target-of-spains-new-tax-on-the-sun/

Poole L 2013 Solar energy battery backup under attack in California? *Renewable Energy World* (7 August 2013) www.renewableenergyworld.com/rea/news/article/2013/08/solar-battery-backup-under-attack-in-california?cmpid=SolarNL-2013-08-0

POST 2015 Energy storage *UK Parliamentary Office of Science and Technology* POSTnote 492 www.parliament.uk/briefing-papers/POST-PN-492/energy-storage

Radford T 2015 Safer battery could spark investment in renewables *Climate Network* (30 September 2015) http://climatenewsnetwork.net/safer-battery-could-spark-investment-in-renewables/

RAE 2012 *Heat: Degrees of Comfort?* (London: Royal Academy of Engineering) www.raeng.org.uk/heat

RMI 2014 *The Economics of Grid Defection* (CO: Rocky Mountain Institute) www.rmi.org/electricity_grid_defection

Rudd A 2015 Statement to the Energy and Climate Change Select Committee (14 July 2015) http://utilityweek.co.uk/news/nuclear-essential-support-for-renewables-until-storage-improves-says-rudd/1153742#.VlMcNoXw-v8

Siegel R 2010 V2G: vehicle to grid will change the whole energy picture *Triple Punbit Blog* (20 July 2010) www.triplepundit.com/2010/07/v2g-vehicle-to-grid-will-change-the-whole-energy-picture/

Smith P 2014 Siemens developing thermal energy storage system *Windpower Monthly* (2 November 2014) www.windpowermonthly.com/article/1320266/siemens-developing-thermal-energy-storage-system

Tesla 2015 Press release *Tesla Powerwall* www.teslamotors.com/powerwall

UBA 2013 *Germany 2050—A Greenhouse Gas-Neutral Country* (Dessau-Roßlau: Umweltbundesamt)

VRB 2012 Vanadium Redox system *Prudent Energy* http://www.pdenergy.com

Williams A 2014 Sand: CSP energy storage solution of the future? *CSP Today* (7 March 2014) http://social.csptoday.com/technology/sand-csp-energy-storage-solution-future

Wilson G, Rennie A and Hall P 2014 Great Britain's energy vectors and transmission level energy storage *Energy Procedia* **62** 619–28

ZCB 2014 *Zero Carbon Britain* (Machynlleth: Centre for Alternative Technology) www.zerocarbonbritain.com/index.php/zcb-latest-report

IOP Publishing

Balancing Green Power

David Elliott

Chapter 4

Grid links to the future

Renewable energy availability varies by time and location, so one way to deal with this is to transmit it from where it is available to where it is needed. That is what is already done with electricity, fossil gas, and sometimes also, over shorter distances, with heat. Renewable electricity, heat and gas can be treated in much the same way. However, in the case of electricity (the focus of this chapter), there may be a possibly helpful variation. Electricity transmission by high voltage AC grid links is widely utilized, but it is not as efficient over long distances as transmission using high voltage DC. HVDC supergrids can be used to balance local variations in renewable supply and local energy demand across whole regions or continents, making use of geographically-defined weather-related variations in availability. In addition to long distance trading of local surpluses, it is also possible to use 'smart grid' systems, perhaps in conjunction with variable 'time of use' pricing arrangements, to manage variable demand and supply, by delaying demand peaks via temporary disconnection of high loads.

4.1 Electricity grids

Some countries have gas grids and local heat distribution networks, both of which can play a role in distributing energy from renewable sources, especially at the local level, most obviously in the case of renewable heat, while, as was noted earlier, gas transmission by pipe has several advantages over power grid transmission. However, in terms of national systems, with electricity having a high utility, its transmission is likely to dominate for the foreseeable future, especially for long-distance transmission. That is the focus of this chapter.

Electricity transmission by overhead power grid cables has become the norm for energy distribution in most industrial countries, linking large centralized power plants, often near fuel sources, to large numbers of consumers via major high-voltage trunk links and then lower-voltage local distribution networks. However, that pattern has begun to break up with the advent of smaller renewable energy systems that use local sources. Even so, since some of these source are variable, the grid system still has a role to play, because it can help balance out local variations in supply and demand, linking up generators and consumers across wide geographical areas. In fact, given the pattern of renewable energy availability, the reach of grids may have to expand to link up across even wider areas. So while some people look to a more locally based and decentralised energy system, it may be that, in parallel, the new energy systems will also rely on long-distance balancing via the grid.

Depending on how long the links would have to be, this would be hard if it used conventional alternating current (AC) high-voltage grid transmission. AC is used primarily because it is relatively easy to convert high-voltage AC to the lower voltages needed by consumers, via local transformer units at the local distribution level. However, there is a penalty, given the energy losses in transmission and in the transformers. The grid cables become hot, and so do the transformers. Up to 10% of the initial energy fed in can be lost over 1000 km transmission and local distribution. By contrast, transmission using high-voltage direct current (DC) is much less wasteful, with around 2–3% lost per 1000 km (Siemens 2015). As figure 4.1 illustrates, the crossover point, where DC wins over AC, is around 700 km.

The downside is that with HVDC it is much harder to tap-off electricity and convert it to what is needed locally, low-voltage AC, at the local distribution level. Expensive DC to AC converters/transformers have to be used, and the switchgear and circuit breakers that are needed are also more expensive than with AC, although new, better, versions are emerging (Harris 2013). Given the high end-of-cable power conversion costs, combined with the lower cost of HVDC transmission, HVDC is usually seen as being best for very long distances, with there only being converters/transformers at each end, feeding local AC distribution networks.

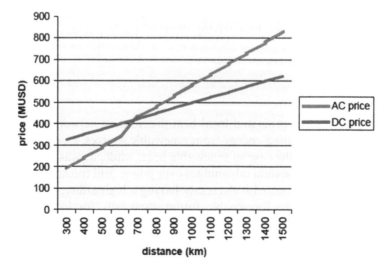

Figure 4.1. AC and HVDC transmission costs for a 2 GW link using overhead lines (Rudervall *et al* 2000). The costs for both will have risen since 2000, but the crossover point will be roughly the same.

4.2 Supergrids

HVDC supergrids, as they are sometimes called, are already quite widely used, for example they link the large hydro projects in central China with the mostly coastal urban centres around 1000 km away. There are also several undersea HVDC links in use around the world, including interconnectors between the UK and mainland Europe. The longest undersea scheme planned so far in the EU is a 730 km 1.4 GW UK–Norway link, to be built by 2020. However, there are also still-to-be-confirmed proposals for an undersea 1.2 GW capacity HVDC link between the UK and Iceland, of around 1000 km in length. It would cost around 4 billion UK pounds.

As that illustrates, supergrid interconnector systems are expensive, especially in the case of undersea links, but the capital cost would be offset by the benefits from the energy transfers that would be enabled, improving energy security and allowing for competition between suppliers across a wider market, potentially cutting costs. Enhanced trading over a wider market is one of the main incentives for supergrids and interconnectors in the EU, as part of its single EU energy market policy and Energy Union programme. The initial stages of that just involve extending existing links, or making new interconnections between the national AC grids. Not all of these need to be HVDV. However, if more energy is to flow, over longer distances, then new HVDC supergrid links will have to be developed. As we shall see, they have begun to emerge as a significant option, opening up a range of opportunities as well as problems (Elliott 2012).

Longer-distance links may be expensive, but there are economic gains in addition to the enhancement of trade over wider areas. Supergrids can link to major renewable resources, and help balance local shortfalls and surpluses across a wider area, which can lead to significant operational cost savings. For example, there would be savings from avoiding the curtailment of local excess renewable output that would otherwise be required, in the absence of storage, at periods of low demand. Supergrids would also give wider access to the large hydro reservoirs that some counties have, opening up more storage possibilities. In addition, judicious exchanges of energy could make use of the (clock) time differentials across regions and the consequent differences in peak demand timings.

However, it may not always be the case that there is surplus electricity to trade. In practice, the price of exchanged electricity would vary, depending on local supply and demand patterns. While some regions may nearly always have a surplus, others may often have to import. One study of the economics of a hypothetical EU supergrid system suggested that by 2030—assuming that the EU is heading for a high renewables mix—many countries would be able to use it to export net excess renewable energy over the year, notably Croatia, Denmark, Portugal and Romania, given their large renewable resources. The only major net importers would be Estonia and Bulgaria (Greenpeace 2014).

Another study focused on the UK and suggested that, assuming around 70% of its electricity came from indigenous renewables by 2050, and given that a large wind capacity would be needed to help ensure there was sufficient output to meet peaks, it would yield a large surplus at other times. Exporting that via supergrid links to continental Europe might enable the UK to earn 15 billion UK pounds per annum, net of imports (Pugwash 2013).

The idea of a pan-EU supergrid network includes the option of links across the Mediterranean sea to North Africa. This would allow imports of electricity from the large solar projects there, and also possibly from wind projects. In the Desertec proposal, around 15–20% of the EU's electricity could be provided in this way, helping to balance-out variations in variable supplies in the EU (Desertec 2015). However, it would not just be an export-based system. Most of the electricity would be used locally, although there might be some bilateral exports across and within the Middle East–North Africa (MENA) region. Indeed, the Desert Power 2050 study produced by the Desertec Industrial Initiative suggested that by 2050 this would be vital, since energy demand would rise in many of the newly emerging economies in the area, possibly by more than in most parts of the EU (DII 2012).

For the MENA countries, trading electricity from their indigenous solar resources would obviously open up new economic and geopolitical issues. At present, some of these countries are economically reliant on oil and/or gas exports. The solar resource in these desert areas is vast, more than enough in principle to meet the needs of the entire MENA area and the EU combined, if fully developed and shared via supergrid networks. Even if it was only partly developed, a solar-driven future could open up new, more sustainable, economic opportunities in the MENA region, while, as the Desert Power 2050 study noted, supergrid links would enhance energy security across the entire EU–MENA area.

Figure 4.2 shows the electricity flows around the fully connected system. The Desert Power 2050 report notes that the Maghreb and Libya are the southern

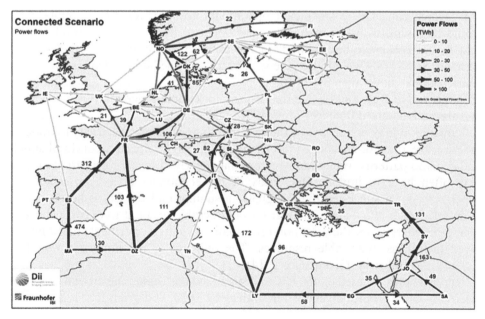

Figure 4.2. Electricity flows around an EU–MENA supergrid system, in TWh. Reproduced with permission from (DII 2012).

'powerhouses' of the region, while Scandinavia, especially Norway, plays the same role in the far north: 'Power flows from the south reach Europe via seven sub-Mediterranean transmission corridors and are then passed on north from Spain, France, Italy and Greece to the UK, BeNeLux countries, Germany, Austria, and the Czech Republic. In the belt from BeNeLux to the Czech Republic, desert power then meets the power flows from Norway via Denmark, Sweden and Poland. An eighth south–north corridor brings power from Egypt and Saudi Arabia on to Turkey'.

Although it was accepted that it would take time and money to deliver a system on this scale, and there are clearly many political and security issues to face in this often unstable part of the world, the Desert Power 2050 report noted that, in its fully connected scenario, the resultant EU–MENA trading would avoid around 33 billion EUR per year in system costs. For the annual electricity exchange of approximately 1110 TWh between the MENA area and Europe that it envisaged, this would amount to a saving of around 30 EUR per MWh. Roughly half of this, the report claimed, would come from the replacement of some of the more expensive renewables used in the EU, with imported solar electricity costs (despite transmission costs and losses) being around 20% lower by 2050, mainly because of the MENA area's high insolation levels. The other half of the savings would be because 'a larger system offers more options to balance the load and the output of solar and wind power plants. Fewer gas peakers for balancing need to be built and less excess production by renewables, i.e. curtailment, occurs'. The report added that the benefits to the MENA countries were also significant: 'it would acquire an export industry for renewable electricity worth up to 63 billion EUR per annum—more than all of the current exports of Egypt and Morocco combined' (DII 2012).

Not all countries would be able to contribute equally. The Desert Power 2050 report identifies three classes of country, based on their renewable resources and their level of energy demand: those with extensive renewables (super producers), renewable-scarce countries (importers) and countries with balanced renewables and demand (balancers). Under the proposed scheme, while no country would be allowed to import more than 30% of its electricity (most already import up to 10%, some more), there would be clear gains for super producers, who could sell their surpluses, and for importers, who could buy in electricity that would cost less than that from their indigenous renewable sources. Morocco, Libya and Algeria are seen as potential super producers (of solar), together with Norway (from hydro). Germany, Italy and, to a lesser degree, France and Turkey, are seen as net importers, because of their high energy demand but relatively low renewable resource base. Egypt, Saudi Arabia and Syria, together with Spain, the UK and Denmark, are seen as the main balancers, with good renewable resources but also high energy demand.

Not every one will agree with these specific assessments and they are inevitably speculative, given the uncertainties about how the energy system will develop in the years up to 2050. However, the overall picture seems clear. A broad EU–MENA supergrid network could in theory benefit all. As the Desert Power 2050 report puts it, 'no single country is dependent on another but instead each country is reliant on the system as a whole'.

There have been similarly ambitious proposals for the Far East, including a supergrid linking solar project in the Gobi desert, and possibly wind projects elsewhere, to demand centres in China, Korea and Japan (Gobitec 2015, CIWG 2013). This would be of particular relevance to Japan, which is hard pressed to support its energy needs without imports and, after the Fukushima nuclear disaster, is keen to use more renewable sources (Elliott 2013).

South Korea has similar problems, with a high population density and high energy imports. In his review of options for this part of the world, using power plants in central Asia, Bent Sorensen notes that 'unfortunately, quite long power transmission lines will be needed to reach South Korea or Japan, at least if passage through North Korea is excluded for reasons of supply security in the importing countries', but, even so, he is hopeful that they could aim for high renewable futures, aided by supergrids (Sorensen 2014)

The potential for renewables in Africa is, if anything, even larger, although there the distances are even more of an issue, and the prospects for supergrids are therefore even more significant. A network of large-scale east–west and north–south HVDC grid links has been proposed, as indicated in box 4.1, although that may be some way off, given Africa's uneven economic development and internal political conflicts. National-level grids are still patchy and often unreliable, and do not reach many rural areas. Local mini-grids may be a sensible intermediate step (IRENA 2013).

The USA could also benefit from improved interconnections across its huge landmass, with there being relatively few links between the large separate regional grids at present. Even so, some of these markets are very large and some links do exist. For example, the PJM Interconnection regional transmission organization covers 13 US states and the District of Columbia, and is the largest competitive wholesale electricity market in the world. However, to extend the market and system integration nationally, there has been talk of 'interstate transmission superhighways' and several studies have indicated that, with better interconnections and/or national supergrids, high renewable contributions could be sustained (Montgomery 2014, MacDonald *et al* 2016).

In terms of specific regional options, there is the ambitious Atlantic Wind Connection offshore HVDC grid proposed for the east coast of the USA, linking up 7 GW of offshore wind projects (AWC 2013). On an even grander scale, there have been proposals for linking up across the whole of North America and beyond, allowing for the balancing of surpluses and shortfalls between Canada, the USA and Mexico, although this would mainly be done by interconnecting their existing grid systems (Sorensen 2014).

There seems to be little prospect of cross-continent links in Australia, given the huge distances and low population density in most of the central region of the country, but there are already links across South America, for example between Argentina, Brazil, Chile, Paraguay and Uruguay, with some of them linking up large hydro projects (Sorensen 2014).

Although at present Russia is not developing (non-hydro) renewables very significantly, the renewable resource is very large, especially for wind across Siberia. It has been suggested that supergrid links could capture this and even export some of the electricity to the EU (Boute and Willems 2012).

> **Box 4.1. Supergrids in Africa.**
>
> In 2012, African Heads of State endorsed the **Programme for Infrastructure Development for Africa** (PIDA), including 15 priority energy projects amounting to a total budget of 40.5 billion US dollars, to be implemented between 2012 and 2020.
>
> As IRENA noted, the project portfolio includes nine hydroelectricity generation projects, four **transmission corridors** and two pipelines, one for oil and the other for gas. The four corridors include:
> - the **north–south transmission link**, from Egypt to South Africa, with branches mostly into East Africa;
> - the **central corridor**, from Angola to South Africa, with branch lines into central and western Africa;
> - a **North African transmission link** from Egypt to Morocco, with links via Libya, Tunisia and Algeria; and
> - the **West African power transmission corridor**, linking Ghana to Senegal, with branches (IRENA 2013).
>
> However, the situation on the ground in Africa may make these ambitious plans difficult to achieve. There are major political and economic hurdles, and it might be argued that, in any case, given that most of the rural population of Africa remains off-grid, the priority should be to get more grassroots renewable energy projects started first, perhaps leading to local mini-grids, which might ultimately feed into national and then international supergrid networks.
>
> The risk with mega schemes is that they will focus on and support large centralised hydro and nuclear power plants, serving major population centers and large industrial complexes, but leaving the bulk of the rural population to rely on local off-grid options.
>
> A counter argument is that a supergrid network could help shift electricity from smaller local projects around the continent and balance the local variations in renewable availability. Most likely, as elsewhere in the world, there will be a mix of the two approaches, although the balance in Africa may be different.

In theory, these huge regional and continent-wide supergrid networks could also be linked up, eventually creating a global supergrid. This offers the tantalising prospect of being able to use daytime solar energy on the night side of the planet. It is a vast, visionary, geopolitically challenging but technically feasible enterprise (Chatzivasileiadis *et al* 2013, Safiuddin 2014).

As gargantuan ideas like that highlight, supergrids offer a very different approach to those favoured by some environmentalists, who would prefer local use of local energy and a decentralised energy system. Decentralists argue that since solar energy is available at some level everywhere, naturally distributed, it makes no sense to reconcentrate it in a few remote areas and then send it back to users far away over expensive grid networks. Small domestic projects certainly can play a role. Some already do, in Germany especially, reducing losses on, and the cost of, local/national grids, with domestic-scale storage also possibly helping, as we saw in chapter 3. We will be looking at that some more below.

However, in principle, the two approaches, local (small scale) and central (large scale), do not have to be mutually exclusive. There is value in using local sources, where available, locally, since this avoids transmission losses and allows for more local control over the usually smaller projects, an important social benefit that is valued by those worried about the power of centralised utilities. Nevertheless, there is also value in using energy sources where they are more concentrated, as in the case of solar in desert areas. Not everyone will have access to good local resources, while some may have too much: trading makes sense. The supergrid just extends that idea more widely. Excess electricity from local sources could feed into the supergrids, with electricity imported when and if there is shortfall, thus helping to balance local variations in supply and demand.

4.3 Assessment of supergrids

Given their potentially very widespread coverage, even if limited to a single continent, the supergrids should be able to help substantially with balancing variable renewables, since they would link up areas with often very different weather regimes. This has certainly been a consensus view.

An early study, TradeWind, a project funded under the EU's Intelligent Energy Europe programme, used European wind power time series data to calculate the effect of geographical aggregation on wind's contribution to overall generation. Its 2020 Medium scenario assumed 200 GW of wind capacity, with a 12% pan-EU wind power market penetration (Tradewind 2009).

Its final report noted that 'the effect of windpower aggregation is the strongest when wind power is shared between all European countries', making cross-EU grid links vital. If no wind energy was exchanged across Europe, the capacity credit (the equivalent firm capacity) was 8%, which corresponded to only 16 GW for the assumed 200 GW installed capacity. But it was noted that 'the wider the countries are geographically distributed, the higher the resulting capacity credit'. So, if Europe was seen as one wind energy production system and wind energy was distributed across many countries according to individual load profiles, then, it said, the capacity credit almost doubled to 14%. This corresponded to about 27 GW of firm power in the system.

A more recent study was carried out by the Fraunhofer Institute for Wind Energy and Energy System Technology, having been commissioned by Agora Energiewende. It found that greater integration of power systems in the Central Western European (CWE) region, i.e. France, Switzerland, Austria, the Benelux countries and Germany, could help to cut costs for balancing weather-related electricity fluctuations significantly, while reducing demands on other parts of the system, and losses due to the need to curtail excess output.

The study assumed that by 2030 around 50% of electricity in the EU would come from renewables, mainly wind and PV. It simulated power production from renewables over a full weather-year. The results showed that networking power systems more closely would reduce the demands for flexibility commonly associated with swings in wind and solar power production, making it possible to balance regional weather differences, and thus the differences in wind power production, on a large scale. Indeed, the effect was so strong that the fluctuation in wind power production across the entire CWE region amounted to only half of all the fluctuation in individual countries. Greater integration also allowed the system to take advantage of national time zone and peak demand differences, and prevented losses from the need to curtail renewable output on very windy or sunny days, since surpluses in one region could offset shortfalls of supply in others. Compared to power systems that are not integrated, such power plant cut-offs could be avoided in 90% of all cases, it was said (Agora 2015).

However, not all studies have agreed that wide geographical spread will aid smoothing significantly. For example, an early study from Poyry consultants found that improved connectivity would only partially alleviate the volatility of increased renewable energy generation. It claimed that 'heavy reinforcement of interconnection doesn't appear to offset the need for very much backup plant'. However, it only looked at northwest Europe, leaving out potentially large inputs from Spain and Eastern Europe, while, oddly, it also ignored Ireland (Poyry 2011).

More recently, a report from the Adam Smith Institute looked at linking up wind operations across the northern European plain, covering Belgium, Holland, Denmark and Germany. It claimed that this 'does little or nothing to mitigate the intermittency of these wind fleets.' It said that for a combined 48.8 GW North European wind system, the actual electricity output only exceeded 90% of the theoretical maximum available for four hours per annum and 80% of the theoretically available output for 65 h per annum, while it would be 20% below that level for 4596 h (27 weeks) per annum, and 10% below for 2164 h (13 weeks) per annum (Aris 2014).

These results reflect the varying load factors in different locations and the fact that, inevitably, outputs will sometimes be much lower than the average, and nearly always less than the maximum possible. The results may also depend of the precise regions covered. Other studies that covered a wider region (including the UK, Spain and eastern Europe) and also looked at the integration of other renewables, not just wind, came to different conclusions. For example, a Greenpeace study, produced jointly with Energynautics, claimed that by 2030 Europe as a whole (including all the 28 eastern and western EU countries, plus Norway and Switzerland) could meet 77% of its electricity needs using a range of renewables, assuming a grid network expansion of 26 000 km. This would reduce the need for surplus output curtailment, seen as otherwise costing around 3 billion EUR per year, by enough to meet the costs of the grid system by 2030 (Greenpeace 2014). Another study, assuming near 100% renewable supply, and using hourly data for solar, wind, and energy demand, found that, with a full renewable continental coverage, and also including links to North Africa, the need for fossil back-up to meet lulls could be halved (Aboumahboub *et al* 2010).

Nevertheless, there are clearly different views, and the debate over the role of supergrids continues, even if much of it seems to depend on the area covered and the range of renewables included, as well as disputes over the quality and interpretation of the data used (see box 4.2).

There is also a wider debate over whether supergrids, whatever their extent and the sources used, are socially and politically acceptable. While small local projects

Box 4.2. The UK debate over smoothing.

In an early exchange of views on intermittency and the potential of interconnectors to smooth out the variations from renewables like wind, two researchers at Imperial College London responded to a critical analysis of the UK situation funded by the arguably oddly named Renewable Energy Foundation, which had concluded that wind power could not be relied on since it varied too much around the UK, with long lulls at times (Oswald *et al* 2008).

The Imperial researchers commented that although the title of the paper was 'Will British weather provide reliable electricity?', its substance concerned the analysis of the impact of variable wind on *individual* conventional generators, which they said was a different issue. What mattered was the impact on the system as whole, as had been the focus of earlier work by the Imperial/UKERC team, reviewing a range of studies which had used a time series statistical approach based on actual plant data, rather than just extrapolations from general Met Office wind data (Gross *et al* 2006).

At root here are conflicting views on and interpretations of data concerning correlations between the output of wind turbines and overall energy supply and demand, and how this is influenced by geographic spread. However, in their response, the Imperial College paper noted that 'Geographical smoothing is not the only reason interconnection assists system operators with the management of intermittency. It also allows the impact of intermittent plants to be shared across a much larger pool of conventional generation and demand variation. This mitigates the effects of intermittency, and reduces the cost of balancing, even if intermittent output is indeed widely correlated across large areas. This is because additional system-balancing requirements are a function of demand prediction errors, overall probability of supply failure and renewable output unpredictability' (Gross and Heptonstall 2008).

Nevertheless, despite further substantial rebuttals (for example, Milborrow 2009), the debate has continued and there are still divergent views about the viability of wind energy smoothing across wider areas (Andrews 2015). With more data from actual wind plant operation now available, it is certainly possible to identify worst-case examples. Thus, looking at Spain, France, Sweden, Denmark, the UK and Germany, Euan Means notes that 'The combined wind capacity of these six countries is 97.9 GW. On occasions the output from this gigantic resource falls below 3 GW, a load of 2.9%. At present and for the foreseeable future the only way to mitigate for wind variability is back-up from other dispatchable power sources. Building interconnectors may provide a marginal partial solution for some of the time but cannot provide a reliable solution at the pan-European scale' (Means 2015).

Certainly, there will be times when there is little wind-derived electricity to trade via interconnectors across most of Europe, although that may not be true when there are more offshore turbines with higher load factors (Morris 2016) and if the wind catchment footprint is also widened to include countries beyond those selected by Means, e.g. from Eastern Europe. Even so, there will also be a need for a range of balancing measures, including storage, flexible back-up and reserve plants, and demand management, as well as interconnectors, along with reliance on a wider range of renewable sources than just wind.

can in theory give more control to local people, supergrids could strengthen the hand of big power utilities. There is also the potential for geopolitical conflicts. For example, depending on the terms of trade, major trading of energy between the EU and Africa could become exploitative and there might be 'land grab' exercises by northern investors keen to obtain access to resources in the south by building solar projects in desert areas. There could also be issues the other way around: the EU would be reliant on imports of energy from North Africa, much as it has been on imports of oil from the Middle East and gas from Russia. There is also the national political issue within the EU that the availability of imports might provide an excuse to avoid developing national renewable resources (Elliott 2012).

Some also worry about security. Supergrids, with long vulnerable cable links across vast areas, might be a target for terrorist attacks. However, a fully developed supergrid network could have a degree of resilience built in. Much like the internet, there could be multiple routes available should one line through be lost for a while, although establishing the necessary redundancy would add to the cost (Stegena *et al* 2012).

That of course raises the issue of who would pay for the supergrid system. The Desertec proposals for North Africa were mainly backed by German energy companies and banks, who would presumably invest in the grid system, and possibly also in the solar projects. They would then expect to earn revenue from energy sales in the EU. However, the host supply countries would also expect a revenue stream and might build and run the solar projects themselves. Some already have, with large concentrated solar power projects now in operation or being built, for example in Egypt, Tunisia and Morocco. They are being developed as national projects, with, so far, the output being just for local use, without any electricity export.

The debate over supergrids has continued in the EU, with the focus being on links within the EU. Some interesting work has been done on the project risks associated with plans for links between EU countries, looking at how cost factors interact with other factors (Torbaghan *et al* 2015). However, utility support for the Desertec MENA initiative has declined. Some key corporate players pulled out of the Desertec Industrial Initiative in 2014 (Deign 2014). There was less impetus for the development of links to North Africa, in part since the EU had by then developed quite substantial renewable resources and arguably did not need imports. That may change, as balancing issues become more important, and it does seem unlikely that the huge solar resource in North Africa will be ignored indefinitely.

For the moment, though, as noted above, what is happening is that large solar projects are being set up locally across the MENA region to serve the local population, without exports being considered as yet. Interconnector links are, however, being developed for balancing and trade within Europe, especially cross-channel links to and from the UK, while there is also the possibility of linking the relatively wind resource-rich north of Germany to the south of the country. Germany is planning a major series of 660 km long HVDC corridors in a 10 billion EUR project (Fairley 2013).

Building new grids on land may be disruptive and there have already been local objections to these plans. There had already been local battles over the 70 km HVDC grid link between Spain and France across the Pyrenees, which were partly resolved by putting some of it underground, at considerable extra cost. That is also now an issue in Germany.

Offshore grids are, of course, less of a problem and they are being developed not just for interconnections, but also to link up offshore wind farms. As offshore wind projects proliferate, for example in the North Sea, they have to be linked to land, and links may be made between those nearer and further out, from both sides of the North Sea, so that the beginnings of a full supergrid network may emerge, incrementally.

4.4 Local power

As noted above, in parallel with supergrid ideas and developments, but actually happening much faster, there has been a rapid expansion of local level generation, including, in Germany, many domestic solar PV projects and also small-scale renewable energy projects being developed by and for local energy co-ops. About 40% of German renewable capacity is now locally owned, by so-called household domestic 'prosumers' and community-based energy co-ops and farmers. This is clearly changing the nature of the German energy market (Debor 2014).

The prosumer movement is focused on PV solar and, as noted in the last chapter, some see the growth of grid-linked domestic rooftop PV, backed up with domestic-scale energy storage and possibly vehicle-to-grid battery charging, playing a role in grid balancing. An Insight Energy study suggested that consumer self-generation, if extended by storage and demand-response measures, 'can reduce the additional integration costs of the integration of PV at high penetration levels (18% of total electricity production) by around 20%', based on its study of the merits of 'self generation and consumption' in the UK, Germany, Belgium and France. This is because demand peaks are reduced and the power supply utility does not have to manage and develop the grid system as much (Insight-E 2015).

While PV is a major element in the new grass-roots movement, it also includes other renewables. For example, local ownership of wind projects is widespread in Denmark, via the many local wind co-ops. In Germany, using solar, wind and biomass, some community-based projects in towns are creating significant local surpluses, which are traded via the grid. So the distinction between local and grid-linked power is becoming blurred, although, clearly, the advent of local generation means less need for centralised power plants. Some studies look to that trend continuing, with a Greenpeace/World Wind Energy Council 2050 global scenario assuming a mix of 70% local and 30% central (Greenpeace 2015).

However, although it obviously favours local-level generation, possibly with local mini-grids, Greenpeace says that 'while a large proportion of global energy in 2050 will be produced by decentralised energy sources, large-scale renewable energy will still be needed for an energy revolution. Large offshore wind farms and concentrating solar power (CSP) plants in the Sunbelt regions of the world will therefore have an important but broader role to play. Offshore wind turbines produce electricity more hours of the day, thereby reducing the need for backup generators, and CSP with storage is dispatchable. Centralized renewable energy will also be needed to provide process heat for industry and desalination (in the case of CSP), to supply increased power demand for the heating and transport sector, and to produce synthetic fuels for the transport sector.'

What that would mean in terms of grid transmission in unclear. There is obviously value in topping up inputs from local sources with inputs from sources elsewhere, via longer-distance grids. If, however, the aim is to minimise the need for grid links and imports, then, to ensure that supplies are sufficient year round, local projects will probably have to be oversized to compensate, or local energy storage facilities will have to be added. Depending on the local context, it would probably make more sense to import balancing supplies, either from nearby communities or, since they may not have enough surplus, from larger projects elsewhere. Greenpeace says that 'in principle, over-sizing local generation locally would reduce the need for large-scale renewable generation elsewhere as well as [for] upgrading the transmission network. In this case the local power system will evolve into a hybrid system that can operate without any outside support. However, making local plants bigger (over-sized) is less economical than installing large-scale renewable energy plants at a regional scale and integrating them into the power system via extended transmission lines'.

They note that the assumed allocation of 70% distributed renewable generation and 30% large-scale renewable generation in their study 'is not based on a detailed technical or economic optimization; in each location, the optimum mix is specific to local conditions. Further detailed studies on regional levels will be needed to better quantify the split between distributed and large-scale renewable generation'.

One reason why you might want to avoid grid trading is to limit the local environmental impacts of transmission links. This has become an issue, given the spread of larger renewable projects like wind farms, which tend to be in remote and often environmentally or aesthetically sensitive areas. The prospect of also having to accept large supergrid links raises the stakes and, as noted earlier, this has also become an issue. However, there may be ways in which the impacts can be reduced. It may be possible for the new HVDC links to simply replace the old AC grid in the same corridors, but if not, HVDC has the advantage that it is possible to bury the cables underground. It is still very expensive, but not as expensive as with AC links. As noted earlier, the latter have higher energy losses per km, with that energy being lost as heat. That can be dissipated easily if they are cables in the air, but if they are underground, cooling becomes an issue. It is less of a problem with lower-loss HVDC.

The German situation is interesting in that while, on the one hand, it seems inevitable that more grids will be needed, on the other hand, such is the impetus for developing renewables, large and small, within Germany, that some think it may not need to bother with imports from other countries, so new long-distance grids may not be needed. It has been argued that, with a national grid network using wind and PV and other renewable inputs (including biomass), along with smart grid demand management and energy storage facilities (including pumped hydro), it may be possible to balance the overall electricity system, without the need for fossil fuel back-up or significant imports. This was demonstrated by the German Kombikraftwerk 'virtual power plant' modeling exercise (Solar Server 2008). In this, wind and PV supplied 78% of the electricity needed, with biogas providing 17% as back-up, along with 5% from storage. On this basis, although local grids would be needed, there would be little need for imports and long-distance grid interchanges.

However, although that degree of national autarchy may be possible within the electricity sector, a report by the German Federal Environment Agency (UBA) said that in order to supply enough electricity to also meet heating and electric vehicle needs, Germany will have to import green electricity from abroad, at a similar scale (but not type) to their fuel import level now. This is in part because their scenario avoids most biomass use (biogas from wastes apart) for environmental reasons. Without it, its own renewable sources would not be sufficient to meet all its energy needs, for electricity, heat and transport (UAB 2014).

Germany's renewable resource is actually quite limited, compared with some other countries (e.g. average wind speeds are low compared with the UK, and its solar input is not high compared with countries further south). Certainly, there are more optimistic longer-term global scenarios which look to renewables being able to expand to cover all energy needs in most countries, even if biomass use is constrained, although the level of imports assumed varies, as does the degree of reliance on large hydro (Jacobson and Delucchi 2011, Delucchi and Jacobson 2013).

Whatever overall route forward is attempted, as can be seen, there will be a range of technical, economic, environmental and social trade-offs when it comes to deciding on the best scale and type of supply and transmission system, with other options, such as local or central storage, also entering into the possible mix. In the next chapter we will be looking at some studies that attempt to identify what optimals there may be. But before we move on to that, it is important to note that, so far, we have only been talking about choices on the supply side and the role of grids in that. What about the demand side? If demand can be reduced or managed, then some of the supply issues, including those concerning balancing and transmission, may be less complex.

4.5 Demand management and smart grids

At the simplest level, energy-saving measures can reduce energy waste and limit demand, so that it is easier to meet it with renewables. Upgrading energy efficiency and avoiding energy waste is a vital priority in any approach to sustainable energy. It will not be easy, but the EU aims to reduce overall energy use by 20% by 2020 and 27% by 2030, while Germany and France have set goals to cut their overall energy demand by 50% by 2050.

However, while it will clearly help to reduce energy use overall, the key issue in terms of balancing is *peak* demand. Meeting demand peaks presents the hardest problem when supplies are variable. Fortunately, there are a variety of ways in which peaks can be reduced.

Perhaps the simplest way to reduce peaks is by variable charging, setting higher prices for energy at peak demand times. That is already done in some countries, with standard higher fixed electricity charges at peak demand times and lower charges at off-peak times. The advent of domestic smart meters may make it possible to do this in an interactive way, with prices varying continually and consumers being alerted to peaks electronically. Few consumers would want the bother of tracking theses changes continually, so as to avoid excessive energy use at peak times. So it is more likely that automated systems would be more popular, perhaps enabling consumers to preset price levels at which various domestic devices would be temporarily disconnected or inoperable, possibly with an 'override' option, in case they still wanted to use them. For example, a tumble dryer may have a warning light saying 'if you use me now if will cost you three times more than if you wait for two hours' or similar.

More draconian approaches are also possible, with selected high energy-using domestic devices simply being cut off by a signal from the grid at peak times, perhaps in exchange for a lower tariff level. Well-insulated trunk freezers units can coast for some hours without power with no significant loss of cooling effect. This type of approach can also be used more widely, for example in the food retail sector, where large freezers are common. Some industries can also happily accept disconnection for short, well-defined periods for some of their activities.

This 'interactive load management' approach would delay energy use to later periods, in effect time-shifting peaks to periods when more energy was available and/or demand lower. It is part of what is sometimes called the 'smart grid' approach, the use of electronic links and controls to manage demand so as to fit supply and improve overall energy system efficiency (see box 4.3).

As the International Energy Agency says in its review of smart grids, part of the aim of smart grid systems is to aid 'informed customer choices about consumption' and enable consumers 'to use energy more prudently' via the provision of information, as with the current generation of smart meters. Along with electricity pricing, it says that smart grids can 'incentivise more sustainable patterns of energy consumption' (IEA 2015).

Box 4.3. Smart grids.

The International Energy Agency's *How2Guide for Smart Grids in Distribution Networks* provides this definition: 'A smart grid is an electricity network system that uses digital technology to monitor and manage the transport of electricity from all generation sources to meet the varying electricity demands of end users.' The IEA says that 'such grids are able to co-ordinate the needs and capabilities of all generators, grid operators, end users and electricity market stakeholders in such a way that they can optimise asset utilisation and operation and, in the process, minimise both costs and environmental impacts while maintaining system reliability, resilience and stability'.

The IEA focuses on smart grid *electricity* integration, but it does note that 'smart grids that use combined heat and power (CHP) have the potential to provide additional benefits of more efficient heat use'. In terms of electricity-based smart grids, it says that they can help move towards sustainable energy infrastructure, 'improving efficiency, facilitating integration of renewable energy sources and providing system resilience, flexibility and security' (IEA 2015).

However, it goes well beyond that. The IEA says that 'ultimately, with greater information flows on how, when and where power is consumed, future energy systems can be designed and operated to more closely match customer's needs.' So it can involve more than simply adjusting demand, it can also involve overall system management, including adjustments on the supply side. The term now widely used to describe this general approach is 'demand-side response' (DSR), although 'demand-side management' (DSM) is still also widely used (see box 4.4).

Whatever the label used, DSM/DSR is not fundamentally about generating energy, although, as box 4.4 illustrates, some plants may be used to balance demand changes in response to consumer demand signals. The main aim is to ensure better balancing of supply and demand. Fine-tuned with price-based smart grid systems, it can help deal with variable renewables, both in terms of shortfalls and surpluses. There would be less need for back-up supplies to meet peaks and, with lower prices when there is excess supply, the need for wasteful curtailment, or expensive storage,

---

Box 4.4. Demand side response.

DSR has been defined as 'a wide range of actions which can be taken at the customer side of the electricity meter in response to particular conditions within the electricity system (such as peak period network congestion or high prices)' (Torriti 2016). However, this definition may be too restrictive, since, as we will see, in practice, some DSR/smart grid approaches can involve adjustments on the supply side, responding to demand signals from consumers.

A review of the potential for demand-side response in the UK electricity sector to 2035 by *Frontier Economics* said that there are a variety of uses for DSR, both on very short timescales (such as for frequency variation/ support) and for longer periods (as in the capacity market, to ensure peak demand can be met). At the same time, by reducing peaks on the transmission and distribution networks, DSR can reduce reinforcement costs and balancing charges and thus the costs faced by utilities and customers.

The report notes that, in fact, most DSR capacity in the UK (about 5 GW according to National Grid) is not currently provided by consumer response, but by distributed generation (smaller-scale conventional, CHP and renewable plants operating flexibly at the local level). It adds that, depending on the future design of market rules, this may persist and expand (to maybe 9 GW by 2035).

The report says that the flexibility this supply-side response provides is likely to be of a greater scale than any individual form of demand-side load shifting (managed interim load disconnection at peak demand times), whether industrial/commercial or domestic. The advent of smart meters and the possible introduction of time-of-use tariffs might nevertheless allow the latter to expand, but in the case of washing machine use, it was not seen as a very effective option compared with most other types of DSR.

However, the report suggested that the UK domestic picture could change when and if electric vehicles, heat pumps and electric storage are widely adopted. Moreover, grid-linked storage (maybe 7 GW by 2035) and enhanced automated voltage control (a ±5% saving option) might also offer significant smart grid potential for load shifting and reduction, respectively (Frontier Economics 2015).

could be reduced. So smart grid/DSR/DSM systems may save money overall, reducing the cost of balancing and avoiding energy waste (Siano 2014).

However, given that smart grids have sometimes been 'over hyped' as being the answer to many energy system problems and, along with home-battery back-up, as being a 'game changer' (Burton 2016), it is perhaps not surprising that some sceptics have doubts about the reality: for example, cynics argue that the main aim of the smart meters and grids is simply to reduce operational and system costs for utilities. That is certainly one motive, but it might lead to savings being passed on as lower consumer prices, and in theory at least, there should also be social and environmental benefits, for example reduced emissions through better matching of supply and demand. Whether these various advantages can and will be captured under current institutional arrangements remains unclear (Hall and Foxon 2014). There may be a need for new, more diverse and decentralized, institutional forms of involvement, including by municipal authorities (Hiteva *et al* 2015)

4.6 System choice

Smart grids and supergrids seem to operate at very different scales. However, there may be potential synergies between these systems, and with some of the other balancing measures we looked at earlier. For example, in theory, smart grid management might be linked to supergrid system management to obtain wider optimals in supply and demand matching, with local and remote storage options also being linked in, as the heat and gas storage and transmission options should be.

In reality, however, there is a long way to go before it will be possible to optimize all the possible components into a coherent overall system. Smart grid/DSR and supergrids are both still in their infancy, as are some of the other options. There are still many uncertainties, potential trade-offs and conflicts. For example, consumers may have negative reactions to smart grids and DSR measures. The UK has been rolling out a much more limited smart meter programme, enabling remote meter reading and easier digital energy-usage readouts for consumers (HMG 2013). However, it has had some problems, due in part to a perception that the aim is simply to reduce costs to utilities, with few benefits accruing to consumers. The digital systems have also evidently proved to be complex, leading to delays (DECC 2014). All new systems can have glitches, but there has certainly been no shortage of adverse reactions and operational problems with the UK smart meter programme (Shannon 2015, Buchanan 2015).

In terms of supergrids, transmission is still expensive and even with HVDC there are losses. Given also that there may not always be surpluses available to import when needed, some say it may be cheaper, easier and more reliable to store any excess energy locally or nationally, rather than exporting it via supergrids, and to use that energy, rather than supergrid imports, to meet shortfalls.

Clearly, then, there are still many choices to be made. The next chapter looks at the potential interactions between some of the key balancing options, including the smart grid and supergrid developments covered in this chapter, and at related system choice, integration and optimisation issues.

> **Chapter summary**
> 1. Grids allow energy to be transferred from where it can best be generated to where it is most needed.
> 2. High-voltage direct current (HVDC) transmission is more efficient over long distances than conventional AC transmission.
> 3. HVDC supergrids can allow local variations in supply and demand to be balanced over long distances.
> 4. Smart grid management systems can allow peaks in energy demand to be delayed until supply is more available.

References

Aboumahboub T, Schaber K, Tzscheutschler P and Hamacher T 2010 Optimization of the utilization of renewable energy sources in the electricity sector *Recent Adv. Energy Environ. Conf.* www.wseas.us/e-library/conferences/2010/Cambridge/EE/EE-29.pdf

Agora 2015 The European power system in 2030—flexibility challenges and integration-benefits *Fraunhofer institute report for agora Energiewende* www.agora-energiewende.org/service/publikationen/publikation/pub-action/show/pub-title/the-european-power-system-in-2030-flexibility-challenges-and-integration-benefits

Andrews R 2015 Wind blowing nowhere *Energy matters* (23 January 2015) http://euanmearns.com/wind-blowing-nowhere/

Aris C 2014 *Wind Power Reassessed: A review of the UK wind resource for electricity generation* (London: Adam Smith Institute/Scientific Alliance) www.adamsmith.org/wp-content/uploads/2014/10/Assessment7.pdf

AWC 2013 *Atlantic wind connection* http://atlanticwindconnection.com/awc-projects/atlantic-wind-connection

Boute A and Willems P 2012 RUSTEC: Greening Europe's energy supply by developing Russia's renewable energy potential *Energy Policy* **51** 618–29

Buchanan B 2015 When amateurs do the job of a professional, the result is smart grids secured by dumb crypto *The Conversation* (14 May 2015) https://theconversation.com/when-amateurs-do-the-job-of-a-professional-the-result-is-smart-grids-secured-by-dumb-crypto-41769?

Burton C 2016 How an energy overhaul will make the national grid redundant *Wired* (25 January 2016) www.wired.co.uk/news/archive/2016-01/25/smart-grids-empower-users

Chatzivasileiadis S, Ernst D and Andersson G 2013 The global grid *Renew. Energy* **57** 372–83

CIWG 2013 A China–East Asia efficient renewable supergrid *China International Working Group* www.ciwg.net/files/74235701.pdf

Debor S 2014 The socio-economic power of renewable energy production cooperatives in Germany: results of an empirical assessment *Wuppertal Institut* http://epub.wupperinst.org/frontdoor/index/index/docId/5364

DECC 2014 *Annual Progress Reports on the Roll-Out Of Smart Meters* (London: Department of Energy and Climate Change) www.gov.uk/government/collections/annual-progress-report-on-the-roll-out-of-smart-meters

Deign J 2014 Desertec: slow death or healthy evolution? *CSP Today* (3 November 2014) http://social.csptoday.com/markets/desertec-slow-death-or-healthy-evolution?

Delucchi M and Jacobson M 2013 Meeting the world's energy needs entirely with wind, water, and solar power *Bull. At. Sci.* **69** 30–40

Desertec 2015 *Desertec Foundation* www.desertec.org/concept/

DII 2012 Desert power 2050: perspectives on a sustainable power system for EUMENA *Desertec Industrial Initiative* http://desertenergy.org/desert-power-2050/

Elliott D 2012 Emergence of European supergrids *Energy Strategy Rev.* **1** 171–3

Elliott D 2013 *Fukushima: Impacts and Implications* (Basingstoke: Palgrave Pivot)

Fairley P 2013 Germany takes the lead in HVDC *IEEE Spectrum (29 April 2013)* http://spectrum.ieee.org/energy/renewables/germany-takes-the-lead-in-hvdc

Frontier Economics 2015 *Future Potential for DSR in GB (Frontier Economics Report)* (London: Department of Energy and Climate Change) www.gov.uk/government/uploads/system/uploads/attachment_data/file/467024/rpt-frontier-DECC_DSR_phase_2_report-rev3-PDF-021015.pdf

Gobitec 2015 *Gobitec Initiative* www.gobitec.org/

Greenpeace 2014 *PowE[R] 2030 (Greenpeace Germany with Energynautics)* www.greenpeace.de/sites/www.greenpeace.de/files/publications/201402-power-grid-report.pdf

Greenpeace 2015 *Energy[R]evolution* 5th edn (Greenpeace/Global Wind Energy Council) www.greenpeace.org/international/Global/international/publications/climate/2015/Energy-Revolution-2015-Full.pdf

Gross R, Heptonstall P, Anderson D, Green T, Leach M and Skea J 2006 *The Costs and Impacts of Intermittency* (London: UK Energy Research Centre) www.ukerc.ac.uk/publications/the-costs-and-impacts-of-intermittency.html

Gross R and Heptonstall P 2008 The costs and impacts of intermittency: an ongoing debate *Energy Policy* **36** 4005–7

Hall S and Foxton T 2014 Values in the smart grid: the co-evolving political economy of smart distribution *Energy Policy* **74** 600–9

Harris S 2013 ABB breakthrough could enable European electrical supergrid *The Engineer* (1 March 2013) www.theengineer.co.uk/energy-and-environment/news/abb-breakthrough-could-enable-european-electrical-supergrid/105665.article

Hiteva R, Foxton F, Nightingale P and Mackerron G 2015 *Response to the Energy and Climate Change Committee's Inquiry into the Future of the UK Electricity Infrastructure* (Sussex Energy Group, University of Sussex) http://blogs.sussex.ac.uk/sussexenergygroup/files/2015/12/Sussex-Energy-Group-Response-to-the-Energy-and-Climate-Change-Committee.pdf

HMG 2013 *Smart Metering Implementation Programme* (London: UK Government) www.gov.uk/government/publications/smart-metering-implementation-programme-information-leaflet

IEA 2015 *How2Guide for Smart Grids in Distribution Networks* (Paris: International Energy Agency) www.iea.org/publications/freepublications/publication/TechnologyRoadmapHow2GuideforSmartGridsinDistributionNetworks.pdf

Insight E 2015 Self-consumption of electricity from renewable sources *Insight Energy EC Briefing* www.insightenergy.org/system/publications/files/000/000/016/original/RREB6_self-consumption_renewable_electricity_final.pdf

IRENA 2013 *Africa's Renewable Future: the Path to Sustainable Growth* (Abu Dhabi: International Renewable Energy Agency) http://www.irena.org/menu/index.aspx?mnu=Subcat&PriMenuID=36&CatID=141&SubcatID=276

Jacobson M and Delucchi M 2011 Providing all global energy with wind, water, and solar power, Part I: Technologies, energy resources, quantities and areas of infrastructure, and materials *Energy Policy* **39** 1154–69

Jacobson M and Delucchi M 2011 Providing all global energy with wind, water, and solar power, Part II: Reliability, system and transmission costs, and policies *Energy Policy* **39** 1170–90

MacDonald A, Clack C, Alexander A, Dunbar A, Wilczak J and Xie Y 2016 Future cost-competitive electricity systems and their impact on US CO_2 emissions *Nat. Clim. Change*

Means E 2015 The wind in Spain *Energy Matters* blog (30 November 2015) http://euanmearns.com/the-wind-in-spain-blows

Milborrow D 2009 *Wind Power: Managing Variability (Greenpeace)* www.greenpeace.org.uk/media/reports/wind-power-managing-variability

Montgomery J 2014 PJM grid operators: we can handle 30 per cent renewable energy integration *Renewable Energy World* (12 March 2014) www.renewableenergyworld.com/rea/news/article/2014/03/pjm-grid-operators-we-can-handle-30-percent-renewable-energy-integration-and-heres-how.html

Morris C 2016 2015: Germany's record wind year *Energy Transitions* blog (21 January 2016) http://energytransition.de/2016/01/2015-germanys-record-wind-year/

Oswald J, Raine M and Ashraf-Ball H 2008 Will British weather provide reliable electricity? *Energy Policy* **36** 3212–25

Poyry 2011 *North European Wind and Solar Intermittency Study* www.poyry.com/news-events/news/groundbreaking-study-impact-wind-and-solar-generation-electricity-markets-north

Pugwash 2013 *Pathways to 2050: Three Possible UK Energy Strategies* (London: British Pugwash) http://britishpugwash.org/pathways-to-2050-three-possible-uk-energy-strategies/

Rudervall R, Charpentier J and Sharma R 2000 High-voltage direct current (HVDC) transmission systems *ABB/World Bank Technology Review paper* http://large.stanford.edu/courses/2010/ph240/hamerly1/docs/energyweek00.pdf

Safuiddin M 2014 Global renewable energy grid project: integrating renewables via HVDC and centralized storage *Renewable Energy World* (19 February 2014) www.renewableenergyworld.com/articles/2014/02/global-renewable-energy-grid-project-integrating-renewables-via-hvdc-and-centralized-storage.html

Shannon L 2015 Just how smart are these energy meters? *Financial Mail on Sunday* (16 May 2015) www.thisismoney.co.uk/money/bills/article-3084432/ERROR-Smart-energy-meters-leave-hundreds-thousands-households-billing-limbo.html

Siano P 2014 Demand response and smart grids—a survey *Renew. Sust. Energy Rev.* **30** 461–78

Siemens 2015 HVDC classic *HVDC Information Site (Siemens)* www.energy.siemens.com/hq/en/power-transmission/hvdc/hvdc-classic.htm

Solar Server 2008 The combined power plant *Solar Server* www.solarserver.com/solarmagazin/anlagejanuar2008_e.html

Sorensen B 2014 *Energy Intermittency* (London: Routledge)

Stegena K, Gilmartin P and Carluccic J 2012 Terrorists versus the Sun: Desertec in North Africa as a case study for assessing risks to energy infrastructure *Risk Manage.* **14** 3–26

Torbagham M, Burrow M and Hunt D 2015 Risk assessment for a UK pan-European supergrid *Int. J. Energy Res.* **39** 1564–78

Torriti J 2016 *Peak Energy Demand and Demand Side Response* (London: Routledge)

Tradewind 2009 Tradewind study of wind power integration in the EU, co-ordinated by the EWEA www.trade-wind.eu/

UBA 2014 *Germany in 2050—a Greenhouse Gas-Neutral Country* (Dessau-Roßlau: German Federal Environment Agency (UBA)) www.umweltbundesamt.de/publikationen/germany-in-2050-a-greenhouse-gas-neutral-country

Chapter 5

System integration

The various options for grid balancing looked at so far all have pros and cons. No one single option seems viable across the board. Studies have attempted to ascertain what the optimal mix in technical, environmental and economic terms might be in a variety of contexts. This chapter reviews some examples of such studies in the EU, and also the USA, covering potential interactions, conflicts and trade-offs amongst candidate options. For example the availability of supergrids would reduce the need for storage and vice versa. It may also be, as one study asserts, that even taken together, storage, smart grids and supergrids may not be enough to avoid continued use of fossil backup or reliance on nuclear energy. The final likely, or best, mix for an optimal integrated system is thus still open to debate and will of course depend on what supply mix emerges.

5.1 System balancing options compared

The previous chapters reviewed the basic options for system balancing, including storage, variable supply, demand management and long-distance supergrid transmission. At present, it is not clear what the best mix will be. Indeed, that will depend on what the emerging energy supply and utilisation system looks like, and that is as yet uncertain. What does seem clear is that it will have a large renewable component.

On that basis, there have been attempts to map out scenarios that show how the various balancing options might help to support variable renewables. They are to some extent shooting at a moving target, since the energy system is evolving rapidly, though unevenly. Unsurprisingly, given this fast changing context, there are divergences of view, with partisan positions sometimes being adopted by supporters of specific options. Moreover, all this is set in a context where there is sometimes opposition to, or least uncertainty over, the expanded use of renewables, with the intermittency issue being one obvious focus.

This chapter reviews the current state of the debate over system balancing, looking at the studies that have emerged so far. The debate can get quite heated. For example, in 2014, Agora Energiewende, a leading German think tank, claimed that post-generation energy storage might not be cost-effective in Germany until the renewables penetration level reached 90% (Colthorpe 2014). DENA, the German Energy Agency, responded angrily: 'Electricity storage facilities are essential for the energy turnaround. Anyone who alleges otherwise is damaging the energy turn-around and, in the end, is risking the supply security in Germany' (DENA 2014).

More nuanced views have emerged from Imperial College London, in a series of studies. The first, in 2012, concluded that by 2050, in the UK context, storage would allow significant savings to be made in generation capacity, interconnection, transmission and distribution networks, and operating costs, providing up to 10 billion UK pounds of added value in a 2050 UK high renewables scenario. However, it was noted that the relative level and share of the savings changed over time and between different assumptions and scenarios. In the high renewables 'grassroots pathway' used by the research team, the value of storage starts out low, but increases markedly towards 2030 and then further towards 2050 (Imperial 2012).

Helpfully, it also looked at what might happen if other balancing options were also adopted. For example, as we have seen, one option for balancing grids in the short term is the use of flexible demand, reducing peaks by time-shifting demand. The Imperial College team said that 'Flexible demand is the most direct competitor to storage and it could reduce the market for storage by 50%'. So with that, not so much storage would be needed. It was similar in the case of supergrid links. They would reduce the need for storage by around 20%. However, this logic can be run the other way around: storage would reduce the need for supergrids. Moreover, all of these options could help reduce curtailment of surpluses. Which option is chosen will depend of the cost and the specifics of the scenario being considered.

Focusing on storage, the Imperial College study found that the value of storage is highest in pathways with a large share of renewables, where storage can deliver significant operational savings through reducing renewable generation curtailment, i.e. when there is excess wind output available and low demand. In addition, storage could lessen the even larger wind curtailment requirement that would result if there is also a significant amount of inflexible nuclear capacity on the grid. Storage was also found to have high value compared to adding carbon capture and storage (CCS): 'adding storage increases the ability of the system to absorb intermittent sources and hence costly CCS plant can be displaced, which leads to very significant savings'.

However, the Imperial College team also noted that, although it can be very useful in some situations, storage is not a magic solution for all grid-balancing problems. It was best used for specific purposes and durations. Crucially, Imperial said that, on the basis of their modeling, while 'a few hours of storage are sufficient to reduce peak demand and thereby capture significant value…the marginal value for storage durations beyond 6 hours reduces sharply to less than £10/kWh year.' So that means short storage cycles, ready for the next demand peak, not long-term grid balancing to deal with long lulls in wind availability.

Clearly, as well as having individual optima, the various options all interact, and these interactions will shape overall system choices, and system economics. It is interesting, therefore, to look at Imperial's subsequent two reports on overall system integration, produced with the NERA consultancy in 2015, by which time the storage options, and indeed the renewables options, had developed significantly, with better estimates of costs being available (Imperial/NERA 2015a and Imperial/NERA 2015b).

5.2 System integration costs

A key finding in the 2015 Imperial/NERA research was that the system integration cost of low-carbon generation technologies will depend significantly on the level of system *flexibility* and that 'very significant cost savings can be made by increasing flexibility'. The flexible options considered include the 'application of more efficient and more flexible generation technologies, energy storage, demand-side response, interconnection' and also reducing the need for various balancing services through 'improved system management and forecasting techniques' (Imperial/NERA 2015b).

So flexibility is seen as a key optimisation principle. The costing details of the overall system, with the various balancing options added, will of course depend on the supply mix, the market situation and operational patterns. The latter are hard to model, especially with regard to the likely supergrid imports and export trades. When there is a shortfall in national supply, prices can rise very dramatically, so trade will prosper. But only if there is something to trade: the north of Europe, although often especially windy in winter, may not always have surpluses, given that it may also be cold, so local demand may be high, and this may sometimes coincide with wind lulls. Similar principles apply to exports to and from pumped hydro storage facilities. For example, there may not be room for extra water in Norway's reservoirs after the spring snow melt, while later in the year they may need most of what they have for their own power generation and will charge more for any exports.

Detailed long-term market balance modeling is not easy, given issues like this, and so far there have been few detailed assessments of possible wide-area energy flow patterns, with the notable exceptions of the study of supergrid interactions produced by the Desertec Industrial Initiative, looked at in chapter 4 (DII 2012), and the more recent EU e-Highway2050 study (2015). However, as they both recognized, translating flows into cost predictions is hard, given uncertainties about future economic and technical developments. There are also many uncertainties about the ease of obtaining 'way leave' rights for long-distance grid links, and political agreement on the integration of national systems.

The situation within individual countries is somewhat easier to model, since supply predictions and generation cost estimates can be made based on national scenarios. In the case of the UK, the Imperial College team produced estimates for wind and PV system integration costs in two scenarios, one assuming 50 g kWh^{-1} carbon reduction, and the other 100 g kWh^{-1} (table 5.1).

On the basis of their estimates of the impact of flexibility, it was concluded that 'enhancing system flexibility reduces system integration cost of renewables by an

Table 5.1. Integration costs (UK pounds per MWh). Data taken from Imperial/NERA 2015b.

Scenario	Wind	Solar PV
100 g kWh^{-1}	6.2–7.6	6.1–9.2
50 g kWh^{-1} (wind dominated)	12.5–15.6	12.1–17.1
50 g kWh^{-1} (solar dominated)	9.5–14.3	26.2–27.6

order of magnitude'. For example, putting that in a comparative context, 'the whole-system cost disadvantage of wind generation against nuclear reduces from circa £14/MWh in a low flexibility system to £1.3/MWh in a fully flexible system achieving 100 g/kWh emission intensity', while the 'whole system cost of solar PV reduces from being £2.3/MWh higher than nuclear to being £10.7/MWh lower than nuclear as the result of improved flexibility' (Imperial/NERA 2015b).

This means more variable renewables can be added to the system while still maintaining system viability, although (impartially) the Imperial team added that it also makes room for more inflexible nuclear. They also said that the alternative for meeting the emission targets would be to add more CCS, which they also see as a flexible option. That is debatable: the economics of plants with CCS, already uncertain, would surely be worsened if they had to vary their output.

However, running all this through their modeling gave a range of possible mixes of supply/balancing capacity, depending on the 2030 emission targets and the degree of flexibility. Sticking with a low-flexibility system, nuclear dominates, but with high flexibility, renewables dominate, with storage making only a small contribution (to flexibility). So the core conclusion from their optimised system studies was that 'flexibility can significantly reduce the integration cost of intermittent renewables, to the point where their whole-system cost makes them a more attractive expansion option than CCS and/or nuclear'.

5.3 Moving beyond LCOE

These optimals are based on total system costs, including full grid balancing, not just individual component levelised costs of energy (LCOE), a holistic approach that the Imperial team see as vital, with system-wide implications. In a parallel paper, Dr Robert Gross from Imperial pointed out that 'demand response, flexible generation, storage and interconnection offer benefits to the system as a whole and building them as if they need to be dedicated to each specific variable renewable installation will result in over-investment. System costs should be charged to generators as cost-effectively as possible, but on the proviso that they are assessed at a system wide level rather than on an assumption that variable renewable installations need to self-balance' (Gross 2015).

Similar views have emerged from other studies. A World Bank study of variable renewable energy (VRE) system optimisation said that it was not helpful just to look at individual balancing or supply options in isolation, focusing just on the lowest cost ones: 'policy, planning and regulatory interventions should be designed to minimize overall system costs subject to meeting performance targets, rather than minimizing the costs of VRE generation alone' (Martinez and Hughes 2015).

A study of optimal approaches to managing high renewable energy mixes in the USA similarly stressed the need to select generation options appropriately, looking at total system costs, not individual component costs, and where possible, choosing supply options with output profiles that complement each other (Becker *et al* 2015).

As can be seen, no one 'ideal mix' emerges from these studies: it is a context- and scenario-dependent question. Unsurprisingly, then, a variety of sometimes differing conclusions have emerged about the merit or otherwise of specific balancing and storage options. For example, a study by researchers at Southampton University, comparing energy storage options and grid interconnections as balancing mechanisms for a 100% renewable UK electricity grid, found that, although all options may be needed, bulk-storing hydrogen in underground caverns was the cheapest option (Alexander *et al* 2015). The potential for this certainly seems to be there. The UK Energy Technologies Institute says that 'the UK has sufficient salt bed resource to provide tens of 'GWe' to the grid on a load following basis from H2 turbines' (ETI 2015a).

By contrast, a Sheffield University study suggested that power-to-gas systems could supply synthetic methane for storage in the gas network (Wilson *et al* 2014). Certainly, as was noted in chapter 2, the gas network can store large amounts of gas, and the power-to-gas idea has good potential in the UK (Qadrdan *et al* 2015)

While there is, arguably, no major difference between these two options (just different gases in different stores), some other studies have differed strongly, and have been critical of the viability of storage and indeed the other balancing options as a means of dealing with renewable variability on a significant scale. For example, the UK Energy Research Partnership (ERP) claims that even 'very significant' storage, demand-side measures and inter-connection would not be sufficient to cope with intermittency in a weather-dependent renewables-based electricity system. It says there would still be a need to have a significant amount of low-carbon firm capacity on the system too, for dark, windless periods. This could, for example, be supplied by nuclear, biomass or fossil fuel plants with CCS (ERP 2015).

5.4 The ERP view

Given that the ERP study is critical of many of the proposals covered in this book, it is worth reviewing it in more detail, and adding some responses. It looks at each of the proposed balancing options in turn and is fairly dismissive of most of them.

Looking at the 'oversizing' approach, the ERP says that it would be counter-productive to just add more wind capacity, since that displaces 'progressively lower carbon plant eventually causing significant levels of curtailment of its own output or that of other zero carbon plant'. While that may be true, you could also say the same for adding extra inflexible nuclear plant. It would potentially force variable wind off the grid.

Power-to-gas conversion and storage of this surplus energy might be a better idea. However, the ERP worries about turn-around efficiency and also says that the scope for batteries, pumped hydro and compressed air or hydrogen storage in salt caverns is limited (it says the largest caverns have 1.4 GWh capacity, so thousands would be needed). Moreover, it says that even if the storage is massively expanded, to 30 GW or more, it would not be sufficient to meet the long-term 8 TWh supply gap produced in their modeling at peak demand times, when the wind is low for long periods.

The ERP says that DSM might help short-term (up to 24 h) by shifting peaks, but 'the 8 TWh gap is not going to be solved through DSM as it represents an average reduction of 15 GW for three weeks. There is little domestic activity that can be delayed that long and the reduction needed exceeds average industrial demand'.

Overall, then, 'neither storage nor DSM seem to be credible solutions to the security of supply issue caused by lulls in renewable output lasting two–three weeks'. However, 'they are likely to have some value on turnaround timescales of hours to a few days on a grid dominated by variable renewables'.

ERP do say that 'Interconnectors could benefit the GB system by connecting it to markets with different weather influences and so take excess generation at times of GB surplus and return carbon free generation at times of low renewable output'. However, they note that 'these interconnected markets would not always be in the right state to do this—for instance when similar weather was being experienced in the neighbouring markets that had installed similar renewable energy technologies. So in effect they would act like storage with an availability that was significantly lower than a physical asset'.

ERP says that the only exception might be for an interconnection to a market such as NordPool, which has significant reservoir hydro, mostly in Norway, which has around 28 GW, used mostly for its domestic needs. About 17 GW of this involves controllable reservoirs, with a total storage capacity of 84 TWh. ERP says that 'a further 5–7 GW could be built without too large of an environmental impact. In theory then 20+ GW of the UK's storage needs could come from Norway, and the 8 TWh needed to fill the low wind gaps could probably be accommodated. In practice though, the UK may find other EU nations also wanting to use NordPool's balancing capabilities and some, unlike the UK, are already connected.'

The ERP concludes that 'interconnection can help, especially to NordPool, but is unlikely to provide a complete solution as other markets compete for the same resources. Furthermore, interconnection does not usually come cheap and a careful examination of the costs involved would be necessary'. But surely *exports* would offset these costs?

Leaving that aside, the ERP insists that 'with the diminishing returns of adding more variable renewables, and the need to cover two–three week periods of low renewable output, a complete decarbonisation is going to need a significant amount of firm low carbon capacity', and it looks at what might work, longer term.

Adding more nuclear (30 GW in one scenario) is seen as beneficial, since it is portrayed as zero carbon. That is not quite true (there are carbon debts from producing the fuel) and, as the ERP admit, it is actually hard to assess the total system impacts from any specific additions, since there are complex interactions, which will change the overall systems operational costs. Certainly, adding nuclear will increase wind curtailment and there may be cost penalties from using basically inflexible nuclear for balancing long lulls: presumably, it would have to be throttled back slightly for the rest of the time.

The ERP look at a system with no nuclear, but with 30 GW of gas/biomass CCS and 56 GW wind. This achieves the emissions targets, but needs fossil back-up. The ERP says that, with a hypothetical 100% renewable system, 'fossil is required to fill 12% of demand,' and there would be 'a significant spill', 8% of generation curtailment. Surely, some of that surplus could be used for power-to-gas conversion and storage, which, along with DSM, would reduce the need for 12% of fossil back-up? However, it could add to the costs.

Trying to identify optimal mixes and system costs in this complex situation is clearly hard. In terms of costs, the ERP notes that 'using DECC's cost estimates, the differences in economic value to the system between the key options examined (nuclear, gas–CCS and onshore wind) are much smaller than the margin of error estimating those costs. Therefore it's difficult to claim any one of these is the optimal solution to progress grid decarbonisation. Furthermore the value to the system is highly dependent on the technology mix on the system, and the effect of diminishing returns reduces the value of all technologies as they are added, but especially so of variable renewables which generate an increasing proportion at times of surplus energy'. So the ERP says that 'using a fixed number (like LCOE) to characterise a technology's economic value is quite unhelpful in these circumstances.'

On that last part, then, they agree with Imperial College and others: LCOE is not a good enough measure. However, as can be seen, on many other issues they adopt quite a conservative view: few of the balancing options are viable as yet, so we may need nuclear and/or fossil CCS.

5.5 Other views on balancing needs and costs

There is nothing wrong with being cautious about new technologies, and the ERP analysis above does have the merit of giving us an idea of how much back-up capacity might needed if these new balancing options do not deliver fully. As noted, it calculates that, even with '100% renewable' electricity supply (with wind and PV meeting most needs most of the time), there might still be a need to have a 12% input from fossil plants to balance the grid during periods of extended low renewable input. That does ignore possible inputs from other less variable renewables, but even so, the 12% figure is a far cry from the assertion sometimes made that there will be a need for 100% *of conventional plant back*-up for variable renewables. For example, in response to a criticism of his paper on wind energy costs for the Global Warming Policy Foundation, Professor Gordon Hughes said that 'to maintain secure reserve margins, each MW of wind generating capacity has to be backed by approximately 1 MW of generating plant which can be run on demand' (Hughes 2012a).

In a similar vein, a report produced for the Adam Smith Institute and the Scientific Alliance, *Wind Power Reassessed: A Review of the UK Wind Resource for Electricity Generation*, looked at a UK mix with 10 GW wind capacity and claimed that 'the model wind fleet would require a conventional generation fleet of equal nameplate capacity to be built and operated alongside it to mitigate the wind fleet deficiencies' (Aris 2014).

We do seem to have moved on from that, and also, mostly, from other unduly pessimistic views on wind power (see box 5.1), with some more realistic estimates emerging of likely integration needs and balancing requirements: there would be no need to install a 100% new back-up system, since most of it already exists and, as the ERP accepted, the other balancing options could reduce how much was needed to some extent.

The issue of costs is, however, more contentious. While the existing system can provide most of the necessary back-up for now, with only marginal extra costs, as more renewables come on the grid, there will be costs in extending and running the system. Given that, as yet, there is only a limited amount of experience with running systems with large inputs from variable renewables, it is hard to come up with reliable estimates for full system and integration costs. They should include direct balancing costs (for the use of extra supply when and if needed), and the cost of providing any extra back-up capacity needed (including storage) and any extra grid links or reinforcement required. Clearly, there will be plenty of room for disagreement about all of these costs, given the uncertainties about what will actually emerge as the energy system changes.

A 2012 study by the Nuclear Energy Agency (NEA) concluded that 'the system costs of variable renewables at the level of the electricity grid increases the total costs of electricity supply by up to one-third, depending on country, technology and penetration levels'. Moreover, while grid-level system costs for dispatchable technologies like nuclear were claimed to be lower than 3 US dollars MWh^{-1}, they were estimated

---

Box 5.1. Wind power reliance.

Critical views on the viability of wind energy have been a regular feature of the debate on renewables energy, with, as noted in chapter 2, some alleging that, at times, there would be no wind available, so full back-up would be needed. The counter-argument that a geographical spread of turbines over a wide area would smooth out local variations, introduced in chapter 1, has been challenged, but, as noted in chapter 4, that view too has been challenged. There will be some smoothing. Moreover, the advent of improved wind forecasting techniques can reduce uncertainty, making it easier to balance the variations with lower reserve margins (Imperial/NERA 2015b).

The debate over wind energy can become quite tortuous, given the involvement of lobby groups critical of too much reliance on renewables. For example, the Global Warming Policy Foundation has published claims that wind turbine performance would fall off very rapidly with age (Hughes 2012b). That was later challenged quite convincingly. It falls off, but slowly (Staffell and Green 2014). Reliance on just wind can certainly lead to local balancing problems within some regions and countries. Ireland, for example, has a small isolated grid, with as yet few interconnectors, and has had to impose limits on how much wind capacity can be used. However, longer term, and taking a wider systems view, with multiple renewable sources (not just wind) and a range of balancing measures, problems like this should not be insurmountable. Nevertheless, there remains the question, raised as we have seen by ERP, of how much fossil fuel back-up may still be needed, an issue we will return to in chapter 7.

to reach up to 40 US dollars MWh^{-1} for onshore wind, up to 45 US dollars MWh^{-1} for offshore wind and up to 80 US dollars MWh^{-1} for solar (NEA 2004).

Some debatable assumptions and data were used to reach these striking conclusions, including arguably low estimates for nuclear back-up/balancing costs and high estimates for renewable back-up/balancing costs. For example, looking at the impact of additional plants, in its 2010 study of the cost of maintaining adequate frequency response via the grid Balancing Services Incentive Scheme (BSUoS) in the UK, National Grid estimated that 'the risk imposed by six additional 1800 MW [nuclear] power stations on the system could increase from £160 m to £319 m'. That works out at about 1.3 US dollars MWh^{-1}, assuming an 80% load factor, not 0.53–0.88 US dollars MWh^{-1}, as quoted for the UK by the NEA for contributions in the range 10–30%. On the renewables side, the NEA's estimate for back-up and balancing for onshore wind in the UK was 21.07 US dollars MWh^{-1}, assuming a 30% contribution, much higher than the 6 UK pounds MWh^{-1} (around 9 US dollars) for a 40% contribution calculated by Milborrow (Milborrow 2009).

However, while the cost of providing balancing may be disputed, it is clearly true that variable renewables will have larger system costs than nuclear, although the presence on the grid of the latter may actually increase both of their costs, since one or the other, nuclear or renewables, or perhaps both, will have to give way when there is excess output over demand. The NEA makes a somewhat different point: 'In systems that currently use nuclear energy, the introduction of variable renewables is likely to lead to an increase in overall carbon emissions due to the use of higher carbon-emitting technologies as back-up.' Basically, given their high and allegedly so far unmet system costs, it would prefer that there not be so much use of renewables. Otherwise, it warns, 'significant changes will be needed to generate the flexibility required for an economically viable coexistence of nuclear energy and renewables in increasingly decarbonised electricity systems'.

That formulation seems to beg the question, what happens if there is no nuclear on the grid? Would that not reduce the potential system conflicts and curtailment problems? The NEA understandably does not see it that way. Instead, it suggests that, unless proper recognition is given to dispatchable sources like nuclear, in the current situation, with renewables not meeting their full system costs, 'dispatchable technologies will increasingly not be replaced as they reach the end of their operating lifetimes, thereby weakening security of supply.' That certainly is a risk, as has been seen in Germany, with gas plants being pushed out of the peak demand market. One of the implications of the NEA analysis seems to be that nuclear will also be constrained (in those countries that use it) and, if it is to be retained, will need extra support to compensate. Unless, that is, renewables are charged the full system cost.

As can be seen, the debate over system costs moves us into wider energy policy issues. However, it should perhaps be noted that some system costs are already reflected in the overall price charged to consumers by suppliers for renewable electricity. Suppliers usually have to pay for using grid links and for their management/development and, for example, in the UK, under the CfD arrangements, generators 'are responsible for any imbalance costs' (DECC 2015). Moreover, the overall price of renewable electricity may fall, as generation costs fall. Wholesale prices certainly have in Germany. It is that trend that may be the main challenge to dispatchable sources, although, as we shall see, as generation costs fall, and as renewable deployment increases, then the system/balancing cost element may be proportionally more significant.

That is a point that emerged from a 2013 analysis from the Potsdam Institute, with a contribution from the Vattenfall utility, which attempted to put the discussion on a somewhat sounder and arguably less partisan basis. In common with Imperial College and the ERP, it argued that LCOE figures are not very helpful and developed a broader 'system LCOE' costing framework, including full system operation and integration costs. Based on a literature review, it concluded that 'at moderate wind shares (~20%) integration costs can be in the same range as generation costs of wind power and conventional plants', and that 'integration costs further increase with growing wind shares'. The results were similar for solar (Ueckerdt *et al* 2013).

Figure 5.1 presents their conclusions for Germany for wind. Note that the short-term balancing costs are seen as surprisingly low and the grid enhancement costs as not much higher. The main costs are the so-called 'profile costs' faced by fossil plants, which have to operate less efficiently and lose some of their peak market share. It is those wider system costs, which fall mainly on utilities, that dominate the Potsdam researchers' assessment of what they call 'integration costs' (including balancing costs). As figure 5.1 shows, at a 40% wind contribution, presumably some way into the future, the overall integration costs (in their usage of the term) are slightly larger than the generation costs.

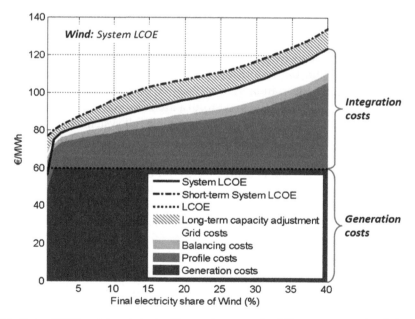

Figure 5.1. System LCOE costs for wind using German data, assuming EU average conventional plant profile: Potsdam Institute results. Reproduced with permission from Ueckerdt *et al* (2013).

The Potsdam report says that this approximate doubling of overall costs with full integration costs added shows why reliance on simple generation LCOE is deceptive, and that this could make it hard for renewables to expand. Even if the renewable plant's LCOE was below conventional energy costs, making it look competitive, in fact, the extra integration costs might negate that. It seems a strong argument. Even though, as renewables expand and develop, generation costs are likely to fall, the integration costs will rise, and so they might dominate and still make renewables less attractive.

However, it may be possible to reduce the integration costs. As we have seen in earlier chapters, there are balancing options that may avoid energy waste (and curtailment) and thus reduce total costs. The reduction in the use of fossil fuel that would follow from the switch to renewables would also help, saving money to offset the system integration/balancing costs. However, that switchover might lead to problems for the utilities, who could be faced with increased costs because of the lower and variable use of their conventional plants; as noted above, these so-called profile costs are very high, according to the Potsdam study. If true, this might be one reason why some utilities are less than keen on renewables (see box 5.2).

Agora Engiewende's study of this issue came up with much lower estimates for integration/system costs. It put the grid/system balancing costs at 5–13 EUR MWh^{-1} (compared to Potsdam's estimates of around 20 EUR MWh^{-1}). Even more significantly, it also looked at the profile costs falling on the utilities, and saw them as being very much lower than the Potsdam estimate of well over 40 EUR MWh^{-1}. Agora's summary said that 'costs resulting from the reduced utilization of conventional plants amount to a maximum of 13 euros per megawatt-hour, with a 50 per cent share of wind and solar energy in Germany'. Moreover, it suggested that 'a flexible adaptation of the power system over the next 20 years can reduce these costs to almost zero.' Under certain circumstances, it claimed that 'the costs of declining conventional power plant utilization could even be negative—renewables would then provide an integration benefit. This could be the case if the reductions in external costs (for example CO_2 costs) outweigh the effect of lower utilization of conventional power plants, or if cheaper renewable energy replaces expensive power from conventional plants' (Agora 2015).

The International Renewable Energy Agency came up with similar views, although somewhat higher figures for total system balancing/integration costs for renewables on a global basis (IRENA 2014). It said that it was hard to do this, given that the figures will depend on the scenarios used, but it very tentatively put the *global* integration/system costs, including the profile costs, 'in the range of 0.035–0.05 US dollars kWh^{-1} with variable renewable penetration of around 40%'. That is significantly less than the 65 EUR MWh^{-1} full integration cost for a 40% wind contribution in Germany shown on the Potsdam chart, which translates to 0.07 US dollars kWh^{-1}. For comparison, the Agora total for Germany translates to a maximum of 0.026 EUR kWh^{-1}.

Like Agora, IRENA notes the wider social and environmental cost benefits of not using fossil fuels: 'when the local and global environmental costs of fossil fuels are taken into account, grid integration costs look considerably less daunting, even with variable renewable sources providing 40% of the power supply'.

Box 5.2. Critique of the Potsdam study.

The Potsdam Institute analysis is interesting but perhaps overstates the problems with renewables. The researchers focus on what they call the 'profile costs' of renewables, incurred by fossil plants, due to their lower usage: 'the average utilization of dispatchable power plants is reduced (reflected in decreasing full-load hours)', resulting in 'inefficient redundancy in the system and higher specific costs compared to the hypothetic situation if wind and solar would not be variable'.

As we have seen, it is true that, with a large input from wind and PV on the grid, fossil plants are less used overall and are sometimes run variably when they are used, all of which makes them less competitive. To an extent, the supply profile they have to adopt is defined by 'must take' feed-in tariffs and other similar priority arrangements, which favour renewables, but it is also a result of simple economics: the renewables are getting cheaper, while the old fossil plants use fossil fuels, which, leaving aside occasional short-term price-fall episodes, are likely to become increasingly scarce and expensive. As green plants expand, the fossil plants are left stranded. They are unable to compete in key markets and lose money. This matters to their owners, whose profits fall, but otherwise it only matters to the extent that the plants are needed for balancing, and they may attract subsidies to that end. However, as we have seen, there are other balancing options, including pumped hydro and other storage systems, demand management to shift peaks, and supergrid imports to deal with supply shortfalls, with these last being balanced by the revenue-earning export of occasional renewable surpluses. They could reduce the balancing costs, but they might not reduce the profile costs. Indeed, they could replace more fossil plant back-up use.

The Potsdam paper looks at some of these options, but, in common with the ERP study looked at above, it does not think they can help very much. Although demand management may be useful, there is not enough pumped hydro storage capacity in Germany to make a lot of difference (only 7.6 GW). However, this is a very static view. Germany is busy working on new storage schemes to rectify this, including new pumped hydro projects and, crucially, wind-to-gas systems, with the green gas then being available for balancing (rather than fossil gas), and it being used, at least initially, in existing (already paid for) gas plants. Some of this may be expensive, but some of it may also pay for itself by avoiding waste and curtailment. We will be looking more at costs and who pays for them in chapter 6. But clearly, to the extent that the Potsdam report reflects utility views, they see the profile costs as staying high, and as potentially doubling the overall costs in the long term.

The International Energy Agency was even more upbeat, claiming that while back-up/grid balancing might add 10–15% to costs with up to 45% penetration, given technological development and higher carbon prices, in time 'the extra system costs of such high shares of variable renewable energy could be brought down to zero' (IEA 2014).

This claim, and Agora's view that the long-term cost might ultimately be negative, may be optimistic, but it is clear that the wider social and environmental costs of using fossil fuel are large and growing, both in terms of health costs due to emissions and poor air quality, and the longer-term impacts of climate change. If those costs are factored in, for example via carbon taxes or carbon pricing, then the cost of the remedial technologies, including the balancing costs, may be less significant.

It will obviously take time to see if the IEA, IRENA, Agora, NEA or the Potsdam Institute are right on the eventual cost of balancing and grid integration and the role that might have in energy system development, but the estimates emerging from some of the other studies we have looked at do look promising. For example, the UK government's advisory Climate Change Committee (CCC), using data from the Imperial College studies looked at earlier, suggested that, for planned 2030 renewable energy supply levels (maybe 40% of total electricity), balancing costs might be around 10 UK pounds MWh^{-1}. That is well under 10% of expected generation costs. However, it added that they would rise at higher levels of renewable penetration, to 15 UK pounds MWh^{-1} for wind at 50 GW and 25 UK pounds MWh^{-1} for PV solar at 40 GW, which translates to around 0.023 US dollars kWh^{-1} and 0.038 US dollars kWh^{-1}, respectively, still well below IRENA's estimate for wind integration/system costs at 40%. Moreover, the CCC claimed that enhanced flexibility of generation and grid balancing services might reduce these costs (CCC 2015).

If the 'external' social and environmental costs are also added in, and new better balancing technologies emerge, some of them actually reducing costs, then it is possible that the final cost will be relatively low and manageable, even with large renewable contributions. However, that will depend on which balancing technologies are used and indeed on which renewables are used.

5.6 The role of heat and balancing the mix

While some clarity seems to have emerged on balancing costs, we still do not have a clear picture of the right mix of supply and balancing systems, with some large issues and choices still having to be resolved. The focus so far in the above has been on electricity and, as we have seen, there is plenty to debate on in that respect. However, as we have also seen, there are heat production and storage options that may be equally important, given that heat is usually the largest single end-use of energy and is easier to store.

The Imperial College team has also had something to say about that. It was quite dismissive of one of the current UK government's favoured options, domestic-scale electric heat pumps, which are meant to replace the use of gas for home heating. They were 'currently more expensive than equivalent gas boilers and, despite their efficiency, heat pumps will only perform effectively and economically in well insulated buildings and may require a change to the internal radiator system, the costs of which must also be considered'. In addition, there were operational issues: 'the performance of air-source heat pumps is particularly dependent on the ambient temperature and falls off rapidly at lower temperatures' (ICEPT 2015).

They were also disruptive to install and had no inherent heat storage capacity. There was the additional problem that their use would put extra stress on the electricity grid. As noted in chapter 2, heat demand is much larger and more variable each day and over the year than current direct electricity demand, so switching heating over to electricity could cause major problems. By contrast, the Imperial team felt that gas heating was a much more flexible and storable option and it was supportive of a shift, where possible, to green gas, including decarbonised syn-gases.

Given that power plants feeding local heat grids could use these gases efficiently, the Imperial team was also keen on community-scaled district heating systems: 'Shared heat systems, due to economies of scale, can be much more efficient in the production of heat, can significantly lower electricity system peak loading', and 'may also more readily incorporate options for storage which can reduce peak demands'.

The heat networks could be fed using a variety of heat sources, including heat from CHP plants, with source changeovers involving 'no disruption to individual users fed by the heat networks.' There were also system integration benefits: 'the amalgamation of district heating with CHP and renewable electricity generation can prove a valuable means of balancing supply and demand across the heat and power sectors'.

The UK Energy Technologies Institute (ETI) has also been looking at heating and energy storage issues. It, too, has warned of the problems of switching from gas to decarbonised electricity for heating: 'a consequence of a low carbon transition is the significant reduction in the flexibility of supply provided by gas heating systems.' And it suggested that 'heat networks could become the system of choice for many UK consumers as they have in a number of Western European cities' (ETI 2015b).

In terms of storage, in some of its 2050 scenarios the ETI sees heat storage linked to district heating networks as offering a major option. Pumped hydro and underground hydrogen storage are ranked much lower, while batteries make an almost negligible contribution, although 'distributed electricity stores' might play a useful role (Coleman 2014, Proctor 2014).

This ranking may seem surprising. Heat stores are rare in the UK at present, although a few new district heating networks are planned. By contrast, pumped hydro is currently the dominant UK storage option, but room for significant expansion, beyond ~50 GWh at most by 2050, is thought to be limited, and the ETI sees storage requirements by 2050 being ten times what pumped hydro can offer. The ETI ranking ignores gas storage: as we have seen, existing gas storage is large, perhaps 250 times larger than the existing electricity storage capacity. And as we have also seen, some of that gas could come from green sources, including green gas provided by power-to-gas conversion.

However, for the moment it is an open field. Which will end up winning? Heat networks would avoid the need to strengthen power distribution links to deal with extra demand for electric heating. But so would the use of gas and, unlike district heating pipes, the gas network already exists.

As can be seen, the technical and system arguments can go round and round, with a range of rankings and prescriptions being offered. The ETI has produced quite an interesting range of UK scenarios, with, in some, nuclear being promoted as a key option (ETI 2015c). However, there are many other UK scenarios, some of which exclude nuclear entirely or avoid new nuclear build in preference for renewables and/or CHP (WWF 2011, Poyry 2011, Pugwash 2013, ZCB 2013, RTP 2015, Greenpeace 2015a).

Most of these scenarios cover heating and transport as well as electricity, although, in some of them, electricity, or hydrogen made from electricity, is used for transport or for heating. See box 5.3 for more detail on two examples, including the balancing arrangements.

The scenarios may differ in detail, for example in relation to their emphasis on heat and gas as opposed to electricity, but for the moment, the simple message from these and other studies is that, in principle, renewables can supply most, if not all, of the electricity needed in the UK, as well as most other energy needs. Moreover, as box 5.3 illustrates, balancing them can also be achieved without major problems.

Many similar studies have emerged for other countries and also with a global perspective, and they have come to similar conclusions. High-renewable scenarios for heat, transport and power are credible and can be balanced (Jacobson and Delucchi 2011, WWF 2013, Sorensen 2014, Greenpeace 2015b).

5.7 Balancing around the world

The focus in the above has been on the UK, but, as we have seen, some of the balancing options are also likely to go ahead quite rapidly around the world, with similar issues and debates emerging. For example, the USA is pressing ahead with renewables and some quite radical projections have emerged about what could be achieved in future, along with assessments of the balancing options.

Box 5.3. The Poyry and Pugwash UK balancing scenarios.

The scenarios referenced in the text are from academic or green NGO sources, with high renewable targets set, but high renewable scenarios have also been produced by and for government agencies. For example, the Poyry consultancy produced a study for the government's advisory Committee on Climate Change, which, along with scenarios with nuclear and carbon capture and storage playing major roles, included a 'max' scenario, with renewables supplying 94% of UK electricity.

In that scenario, balancing was achieved by having a large renewable surplus, with 264 GW installed (including 189 GW of wind plant), supplying 674 TWh per annum, along with 4 GW of bulk storage (20 TWh capacity), demand management (with price signals yielding around 1 GW of reduction), and 12 TWh per annum net of imports over exports, all topped up by 34 GW of gas plants, 21 GW of them being peaker plants, supplying 38 TWh per annum. That mix was able to meet the projected overall electricity demand, which was put at 550 TWh by 2050, while also providing 120 TWh for hydrogen production by electrolysis for use in vehicles. It was also able to meet peak demands, even during low wind/solar periods (Poyry 2011).

Most of the other UK scenarios referenced above assume that energy demand can be reduced, so that less renewable overcapacity is needed. For example, the Pugwash high renewables scenario, which I helped work on, assumed a 40% reduction in demand by 2050 and only had 106 GW of wind capacity, although similar levels (to Poyry) of other renewables. It, too, was able to meet demand reliably, including peaks, using a mixture of mainly non-fossil balancing techniques including storage, demand-side management and interconnectors. It also met some heat demand directly using solar and biomass. Some of the implications of this study are explored in chapter 7: for example, it suggests that, with more renewables and power-to-gas extensions, the need for fossil back-up could be minimal.

In a recent study, Jacobson *et al* at Stanford University produced advanced overviews of potential 100% renewable energy supply mixes covering 50 US states, including analysis of integration and balancing issues (Jacobson *et al* 2015a). They said that the overall mix, supplying the full range of loads with electricity, would deliver 30.9% of the energy needs of the 50 states from onshore wind, 19.1% from offshore wind, and 30.7% from utility-scale PV, with rooftop PV, CSP with storage, geothermal power, wave tidal and hydro power making up the rest. However, they added, based on a parallel grid integration study, there would also be a need for additional capacity, beyond that needed for annual loads, for peaking and grid stability, mostly based on CSP with storage (see box 5.4).

There have also been studies of the impacts of specific technologies. In the case of wind power, the first of the new renewables to be developed widely in the US, improved wind forecasting has often been highlighted as a key practical issue, since this can aid the scheduling of reserve capacity and grid management, but grid upgrades and storage have also been seen as important. In the case of solar PV, which is now expanding rapidly, there have been some interesting studies of the implications for grid management and balancing. For example, a study by Argonne National Labs found that with high PV contributions (they used 17% of total generation in the Arizona region as a basis for the study), 'curtailments of renewable energy reach a very high level (17.8% of the renewable potential) and that satisfying balancing reserve requirements is challenging in a few hours of the year'. However,

---

Box 5.4. Balancing in the USA.

In their state-by-state study of potential renewable energy development option across the USA by 2050, focused on the use of WWS, Jacobson *et al* looked at the system balancing issues in some detail.

As they report 'Wind and solar time-series were derived from 3D global model simulations that accounted for extreme events and competition among wind turbines for kinetic energy and the feedback of extracted solar radiation to roof and surface temperatures. Solutions were obtained by prioritizing storage for excess heat (in soil and water) and electricity (in ice, water, phase-change material tied to CSP, pumped hydro, and hydrogen), using hydropower only as a last resort, and using demand response to shave periods of excess demand over supply.'

Interestingly, 'No stationary storage batteries, biomass, nuclear power, or natural gas were needed. Frequency regulation of the grid was provided by ramping up/down hydropower, stored CSP or pumped hydro; ramping down other WWS generators and storing the electricity in heat, cold, or hydrogen instead of curtailment; and using demand response'.

In conclusion, they noted that 'Multiple low-cost stable solutions to the grid integration problem across the 48 contiguous US states were obtained, suggesting that maintaining grid reliability upon 100% conversion to WWS in that region is a solvable problem', adding that they found that US-averaged LCOE, including storage, transmission, distribution, maintenance costs and array losses, 'was ~10.6 US cents kWh^{-1} for electricity and ~11.4 US cents kWh^{-1} for all energy in 2013 $' (Jacobson *et al* 2015b).

'with increased flexibility the estimated integration costs vary between 1.0 and 4.4 US dollars MWh^{-1} PV in the high PV scenario. Increased flexibility also reduces the curtailment of renewables to between 0.9% and 9.1% of the renewable potential, indicating that the increased system flexibility makes it much easier to absorb high solar PV penetration levels' (Mills *et al* 2013).

There have also been studies that look at local/regional balancing issues with large renewable penetrations. For example, one that looked at California found that 'the largest increase in the value of wind at high penetration levels comes from increased geographic diversity', while 'the largest increase in the value of PV at high penetration levels comes from assuming that low-cost bulk power storage is an investment option' (Mills and Wiser 2015).

Similar conclusions were reached in a four-year RenewElec project overseen by Carnegie Mellon University, the University of Vermont and the Van Ness Feldman environmental law firm. It looked at the options for medium-scale (20–30%) integration of variable electric power generation in the USA as a whole. It concluded that there was value in extra local/regional interconnections. For example, it found that electrically combining the output of several wind plants in a region can reduce variability in the aggregate electricity output, although the extent to which this occurs depends on the timescale involved. The impacts from short-term variations (~1 h) can be cut by 95%, while 12 h variations can only be cut by 50%. The benefits also fall off sharply as more plants are added in the same area; there are diminishing returns. However, expanding the area and aggregating wind power generation over a large geographical region was found to be beneficial to reducing variability, although making the links may cost more (in the US context) than building new CCGT gas plants. Hence major new grid links were not seen as the best option (Apt and Jaramillo 2014).

A somewhat different conclusion on grids was reached in a NOAA study by researchers at the University of Colorado, Boulder. They developed a cost-optimising model of US energy system development running up to 2030. On the basis of projected solar and wind cost reductions, it selected supergrid links, enabling variable solar and wind to expand significantly. Carbon emissions from the power sector were reduced by up to 80%, with this achieved 'by moving away from a regionally divided electricity sector to a national system enabled by high-voltage direct-current transmission.' (MacDonald *et al* 2016). The study found that investing in efficient long-distance transmission was the key to keeping costs low, with, in the high renewables scenario, electricity costs put at 10 US cents kWh^{-1}, using current technologies and without electrical storage (NOAA 2016).

The same issues are emerging elsewhere. For example, with renewables expanding rapidly in China, balancing issues are becoming urgent there, especially in relation to wind energy and curtailment problems. Given the long distances between the bulk of the wind resource and the main energy demand in the major cities, transmission issues have come to the fore. We will be looking at that in the next chapter: supergrids, already used to link large hydro projects, are one possible answer.

Similar issues will emerge elsewhere as renewables expand, especially in Africa. Clearly, there are plenty of often complex technical balancing challenges ahead, around the world. However, the overall pattern of likely development seems clear. The use of fossil plants for balancing will continue and may well expand, though perhaps increasingly using green gas. Storage looks set to expand rapidly at all scales, while heat supply, heat networks and CHP may well begin to be taken more seriously, but demand-side management may only make a more gradual entrance. New supergrid-type interconnectors are already planned, for example, across the North Sea, and there may be progress on further link-ups between national energy transmission networks across the EU. This will mainly involve AC links, but HVDC supergrids may follow, and in time spread to North Africa. Similar long-distance supergrid projects could also emerge elsewhere in the world.

This may all happen in an *ad hoc* and incremental way, driven by varying local/national economic and operational concerns and technical innovations. However, there is also merit in a more coherent approach. This implies some degree of co-ordination. Indeed, in some cases, it will be vital. For example, smart grids and heat networks need careful local planning and supergrids open up wide-ranging planning and co-ordination issues.

More generally, there are broader strategic policy issues to be faced in deciding which of the various balancing approaches to promote. While it would be wise to seek technical optima as far as possible, the final mix that emerges will probably depend more on the economics, and the vagaries of politics. The final chapter therefore looks at how the process of choice, and support for emergent choices, is being carried out in the UK and elsewhere.

The ETI says, perhaps a little wistfully, that 'the interaction between energy vectors seems set to increase in future systems with a need to integrate the operation of the gas, electricity and heat sectors', but 'there is currently no owner for the holistic view of integrated electricity, gas and heat systems' (ETI 2015b). However, choices are being made. So the next chapter looks at how these issues are being faced.

Chapter summary

1. No one balancing system is necessarily the best—their suitability depends on the context and in most cases a mix is likely to be needed.
2. There can be conflicts and trade-offs between possible balancing options: some may pre-empt the need for others.
3. There are divergent views on what the optimal mix might be, on the role of heat, gas and electricity in the mix, and on whether the various non-fossil options will be sufficient to balance high levels of variable renewables.
4. Estimates of the likely costs of balancing large renewable inputs vary, but some look relatively low, depending on technological developments and the value attributed to carbon abatement.

References

Agora 2015 *The Integration Costs of Wind and Solar Power* (Berlin: Agora Energiewende) www.agora-energiewende.de/fileadmin/Projekte/2014/integrationskosten-wind-pv/Agora_Integration_Cost_Wind_PV_web.pdf

Alexander M, James P and Richardson N 2015 Energy storage against interconnection as a balancing mechanism for a 100% renewable UK electricity grid *IET Renew. Power Gen.* **9** 131–41

Apt J and Jaramillo P 2014 *Variable Renewable Energy and the Electricity Grid* (London: Routledge)

Aris C 2014 *Wind Power Reassessed: A Review of the UK Wind Resource for Electricity Generation* (London: Adam Smith Institute/Scientific Alliance) www.adamsmith.org/wp-content/uploads/2014/10/Assessment7.pdf

Becker S, Frew B, Andresen G, Jacobson M, Schramm S and Greiner M 2015 Renewable build-up pathways for the US: generation costs are not system costs *Energy* **81** 437–45

CCC 2015 Power sector scenarios for the fifth carbon budget *Advisory Committee on Climate Change* (London: UK Government) www.theccc.org.uk/2015/10/22/new-low-carbon-electricity-generation-is-cost-effective-option-for-uk-power-sector-investment-in-2020s-and-beyond/

Coleman J 2014 Energy storage—a global challenge and a global prize *Energy Technologies Institute presentation to a clean energy conference organised by The Engineer* www.eti.co.uk/wp-content/uploads/2014/06/Energy-Storage-The-Engineer-Conference-v3-CLEAN.pdf

Colthorpe A 2014 Report challenges short-term role of storage in Germany's energy transition *PV Tech* (22 September 2014) www.pv-tech.org/news/energy_storage_not_needed_in_germany_until_nation_hits_90_renewables_penetr

DECC 2015 UK government response to the Energy Market Design consultation (London: Department of Energy and Climate Change) www.gov.uk/government/publications/uks-response-to-european-commission-consultation-on-energy-market-design

DENA 2014 Dena calls for rapid expansion of electricity storage facilities *Press Release* (7 October 2014) www.dena.de/en/press-releases/pressemitteilungen/dena-fordert-stromspeicher-muessen-zuegig-ausgebaut-werden.html

DII 2012 Desert power 2050: perspectives on a sustainable power system for EUMENA *Desertec Industrial Initiative* http://desertenergy.org/desert-power-2050/

e-Highway 2050 2015 *Europe's future secure and sustainable electricity infrastructure: e-Highway2050 project results (RTE, Paris)* www.e-highway2050.eu/results

ERP 2015 *Managing Flexibility whilst Decarbonising the GB Electricity System* (London: Energy Research Partnership) http://erpuk.org/wp-content/uploads/2015/08/ERP-Flex-Man-Full-Report.pdf

ETI 2015a *Hydrogen Storage Insight Report* (Loughborough: Energy Technologies Institute) www.eti.co.uk/storing-hydrogen-underground-in-salt-caverns-and-converting-it-into-a-reliable-affordable-flexible-power-source-could-help-meet-future-uk-peak-energy-demands-according-to-the-eti/

ETI 2015b *Heat Insight: Smart Systems and Heat—Decarbonising Heat for UK Homes* (Birmingham: Energy Technologies Institute) www.eti.co.uk/wp-content/uploads/2015/03/Smart-Systems-and-Heat-Decarbonising-Heat-for-UK-Homes-.pdf

ETI 2015c *UK Scenarios For a Low Carbon Energy System Transition* (Loughborough: Energy Technologies Institute) www.eti.co.uk/wp-content/uploads/2015/02/Options-Choices-Actions-Hyperlinked-Version-for-Digital.pdf

Greenpeace 2015a *2030 Energy Scenario Demand Energy Equity Report (Greenpeace UK)* www.demandenergyequality.org/2030-energy-scenario.html

Greenpeace 2015b *Energy[R]evolution* 5th edn (Greenpeace/Global Wind Energy Council) www.greenpeace.org/international/Global/international/publications/climate/2015/Energy-Revolution-2015-Full.pdf

Gross R 2015 Driving innovation through continuity in UK energy policy *Briefing paper* (London: Centre for Energy Policy and Technology, Imperial College London) https://workspace.imperial.ac.uk/icept/Public/innovation%20and%20continuity%20in%20UK%20policy%20%283%29.pdf

Hughes G 2012a *Response to Goodall and Lynas, Global Warming Policy Forum* (28 September 2012) www.thegwpf.com/gordon-hughes-response-to-goodall-lynas/

Hughes G 2012b *The Performance of Wind Farms in UK and Denmark* (London: Global Warming Policy Foundation) www.ref.org.uk/publications/280-analysis-of-wind-farm-performance-in-uk-and-denmark

ICEPT 2015 Energy system crossroads—time for decisions *ICEPT Discussion paper* (London: Imperial College London) https://workspace.imperial.ac.uk/icept/Public/Energy%20System%20Crossroads.pdf

IEA 2014 *The Power of Transformation—Wind, Sun and the Economics of Flexible Power Systems* (Paris: International Energy Agency)

Imperial 2012 *Strategic Assessment of the Role and Value of Energy Storage Systems in the UK* (London: Low Carbon Energy Future Energy Futures Lab, Imperial College London) www.carbontrust.com/resources/reports/technology/energy-storage-systems-strategic-assessment-role-and-value

Imperial/NERA 2015a System integration costs for alternative low carbon generation technologies—policy implications *Imperial College London/NERA Consultants Report for the Committee on Climate Change* www.theccc.org.uk/publication/system-integration-costs-for-alternative-low-carbon-generation-technologies-policy-implications/

Imperial/NERA 2015b Value of flexibility in a decarbonised grid and system externalities of low-carbon generation technologies *Imperial College London/NERA Consultants Report for the Committee on Climate Change* www.theccc.org.uk/publication/value-of-flexibility-in-a-decarbonised-grid-and-system-externalities-of-low-carbon-generation-technologies/

IRENA 2014 *Renewable Power Generation Costs in 2014* (Abu Dhabi: International Renewable Energy Agency) www.irena.org/menu/index.aspx?mnu=Subcat&PriMenuID=36&CatID=141&SubcatID=494

Jacobson M and Delucchi M 2011 Providing all global energy with wind, water, and solar power, Part I: Technologies, energy resources, quantities and areas of infrastructure, and materials *Energy Policy* **39** 1154–69

Jacobson M and Delucchi M 2011 Providing all global energy with wind, water, and solar power, Part II: Reliability, system and transmission costs, and policies *Energy Policy* **39** 1170–90

Jacobson M *et al* 2015a 100% clean and renewable wind, water, sunlight (WWS) all-sector energy roadmaps for the 50 United States *Energy Environ. Sci.* **8** 2093–117

Jacobson M *et al* 2015b 100% clean and renewable wind, water, and sunlight (WWS) all-sector energy roadmaps for 139 countries of the world (Stanford, CA: Stanford University) http://web.stanford.edu/group/efmh/jacobson/Articles/I/CountriesWWS.pdf

MacDonald A, Clack C, Alexander A, Dunbar A, Wilczak J and Xie Y 2016 Future cost-competitive electricity systems and their impact on US CO_2 emissions *Nat. Clim. Change* doi:10.1038/nclimate2921

Martinez S and Hughes W 2015 Bringing variable renewable energy up to scale: options for grid integration using natural gas and energy storage *World Bank* http://documents.worldbank.org/curated/en/2015/02/24141471/bringing-variable-renewable-energy-up-scale-options-grid-integration-using-natural-gas-energy-storage

Milborrow D 2009 Quantifying the impacts of wind variability *Energy* **162** 105–11

Mills A, Botterud A, Wu J, Zhou Z, Hodge B-M and Heaney M 2013 *Integrating Solar PV in Utility System Operations* (Argonne, IL: Argonne National Laboratory) https://emp.lbl.gov/sites/all/files/lbnl-6525e.pdf

Mills A and Wiser R 2015 Strategies to mitigate declines in the economic value of wind and solar at high penetration in California *Appl. Energy* **147** 269–78

NEA 2012 *Nuclear Energy and Renewables: System Effects in Low-Carbon Electricity Systems* (Paris: Nuclear Energy Agency/OECD) www.oecd-nea.org/press/2012/2012-08.html

NOAA 2016 Rapid, affordable energy transformation possible *Press Release* (26 January 2015) (Washington, DC: National Oceanic and Atmospheric Administration, US Department of Commerce) www.noaanews.noaa.gov/stories2016/012516-rapid-affordable-energy-transformation-possible.html

Pöyry 2011 Analysing technical constraints on renewable generation to 2050 *Report for the Committee on Climate Change* www.theccc.org.uk/archive/aws/Renewables%20Review/232_Report_Analysing%20the%20technical%20constraints%20on%20renewable%20generation_v8_0.pdf

Proctor P 2014 Energy storage: a potential game changer and enabler for meeting our future energy needs? *Energy Technologies Institute Presentation to the All Energy Conference, Aberdeen* www.eti.co.uk/wp-content/uploads/2014/05/All-Energy-Storage-May-14-Phil-Proctor.pdf

Pugwash 2013 *Pathways to 2050: Three Possible UK Energy Strategies* (London: British Pugwash) http://britishpugwash.org/pathways-to-2050-three-possible-uk-energy-strategies/

Qadrdan M, Abeysekera M, Chaudry M, Wu J and Jenkins N 2015 Role of power-to-gas in an integrated gas and electricity system in Great Britain *Int. J. Hydrog. Energy* **40** 5763–75

RTP 2015 *Distributing Power: a Transition to a Civic Energy Future* (Bath: Realising Transition Pathways Research Consortium, University of Bath) www.realisingtransitionpathways.org.uk/realisingtransitionpathways/news/distributing_power.html

Sorensen B 2014 *Energy Intermittency* (London: Routledge)

Staffell I and Green R 2014 How does wind farm performance decline with age? *Renew. Energy* **66** 775–86

Ueckerdt F, Hirth L, Luderer G and Edenhofer O 2013 System LCOE: what are the costs of variable renewables? *Energy* **63** 61–75

Wilson G, Rennie A and Hall P 2014 Great Britain's energy vectors and transmission level energy storage *Energy Procedia* **62** 619–28

WWF 2011 *Positive Energy: How Renewable Electricity Can Transform the UK by 2030* (London: World Wide Fund for Nature, London) http://assets.wwf.org.uk/downloads/positive_energy_final_designed.pdf

WWF 2013 *100% Renewable Energy by 2050 for India* (New Delhi: World Wildlife Fund for Nature/TERI) www.wwfindia.org/?10261/100-Renewable-Energy-by-2050-for-India

ZCB 2013 *Zero Carbon Britain* (Machynlleth: Centre for Alternative Technology) www.zerocarbonbritain.com/

Chapter 6

Making changes

As the use of renewables expands, it will be necessary to provide more grid balancing facilities. This chapter looks at how this type of development has been promoted, focusing initially on the different market-based approaches being adopted in the UK and Germany. It seems clear that as renewables develop, the emphasis will move away from large base-load plants to more flexible energy systems both for supply and demand management. That transition is currently uneven, in that, for example, unlike Germany, the UK is still supporting large centralised nuclear projects. Whether that approach will avoid balancing problems is far from clear. But the flexibility concept would suggest that it will not. More likely a shift to new more decentralised supply technology will become the norm and will require a shift to new balancing technology. That means developing ways to ensure that the necessary balancing services are available, with some examples of current practice around the world being looked at, including in the USA and China, along with some of the linked implications for institutional change.

6.1 Renewables and grid balancing in the EU

In *Powering the Future*, a 2009 report on UK energy options by the engineering consultancy Parsons Brinckerhoff, it was claimed that 'the current mix of generating plant will be unable to ensure reliable electricity supply with significantly more than 10 GW of wind capacity' (PB 2009).

The UK now has over 13 GW of wind capacity, with much more planned, and balancing does not yet seem to be a major problem. This is also the case for Denmark, which now gets 43% of its electricity annually from wind. Balancing may, of course, become more of an issue as renewables expand further, especially for large energy-using countries like the UK that do not have many grid links to other countries at present. Denmark is a small country that is able to link up to a range of neighbours, so that, for example, it can use hydro in Norway for balancing.

Germany, by contrast, is a large and energy-hungry country, but it is also able to link up to neighbouring power systems, although that can be problematic, given the size of its energy system.

As renewables expand, additional balancing measures will be required by these and other countries. This chapter looks at how new and existing balancing measures are being made available and in particular at the support and promotion mechanisms being used. As we have seen, the measures being taken to ensure balancing vary. Interconnector links are only one option and in some countries they may be less favoured than in others. There can be short-term practical problems. For example, Germany is having problems getting local agreements for new major north–south grids and meanwhile is also having problems with routing some its electricity through neighbouring countries' grid systems. They claim that it is disruptive and that they are not being paid sufficiently for this service (Oroschakoff 2015).

Energy storage is still relatively underdeveloped, although Germany is pushing ahead with new schemes for pumped storage and, like Austria, is linking up some existing hydro projects to provide pumped storage services. Power-to-gas projects are also spreading, but so far these are relatively small, as are the various utility-scale battery projects. Some visionary schemes are being considered, such as Denmark's Green Power Island offshore reservoir. This would use electricity from a nearby offshore wind farm to pump water out of the impounded reservoir and then generate electricity when needed by letting the water back in through hydro-type turbines. Some tidal generation might also be possible (Green Power Island 2015). There is also an interesting EU programme for smart grids, including local field tests of demand-response systems on a Danish island (Ecogrid 2015).

However, apart from large and small, often longer-term, projects like these, the main emphasis has been on using conventional plants for back-up. Although some of this capacity may already exist, there may be a need for some additional effort to ensure that the necessary capacity, new and old, is available in future. For example, as was noted earlier, the UK has set up a capacity market to provide support, via a consumer levy, for projects that can help balance the grid. This was expected to put about 11 UK pounds per annum on typical household energy bills. In its first competitive auction round in 2014, nearly 50 GW of capacity was contracted to be available for this purpose. Most of it was existing fossil capacity, including 22 GW of combined cycle gas turbines and around 15 GW of other fossil plants, including CHP. Only 2.7 GW of storage capacity was included and just 174 MW of DSR. Controversially, nearly 8 GW of existing nuclear capacity was also included, but no interconnector projects were (National Grid 2015).

There were criticisms of this outcome as supporting the energy status quo, with few new projects being backed. It seemed more like an attempt to provide an extra subsidy for existing plants, while almost completely ignoring new balancing options like DSR, which accounted for only 0.35% of the total projects supported. Certainly, some have seen capacity markets as being problematic, given this trend (Harrison 2015).

A policy announcement in 2015 did little to dispel fears that the UK capacity market was mainly seen as a top-up device for supporting extra conventional capacity, not as a way to support the development of new smart grid, storage and demand-side response balancing systems. The Energy Secretary said that, since 'in the next 10 years, it's imperative that we get new gas-fired power stations built', the government was 'consulting on how to improve the Capacity Market' so as to 'ensure it delivers the gas we need' (Rudd 2015).

Interestingly, Germany seems to be adopting a very different approach, and has taken the view that capacity markets will lock existing fossil plants into a back-up role. Rather than providing a subsidy for selected projects, it argues that, suitably structured, the market will ensure that the right capacity, including new balancing capacity, is available.

6.2 The German approach

A 2015 federal government white paper on the proposed new 'electricity market 2.0' said that 'Capacity markets would have negative effects on the market and system integration of renewable energy, would increase carbon emissions, would not be ecological and would delay the energy transition and the renewal of the power plant fleet. They would weaken the price signals of the electricity markets and thus impede the necessary flexibilisation of the electricity system'.

Instead, they said that the proposed electricity market 2.0 'does not require any intervention in the market mechanism and is thus less susceptible to regulatory failure. A competitive system will bring out the cheapest solutions for the integration of renewable energy sources. As a result, the electricity market 2.0 creates incentives for new fields of business and sustainable solutions' (BMWi 2015).

Will market pressures really be sufficient to create the necessary capacity? In a preface to the white paper, the minister noted that 'we do not necessarily need more power stations, but rather flexible capacity. Flexibility is the answer to the weather-dependent renewable energy sources. By introducing the electricity market 2.0, we are permitting fair competition between all flexibility options. These include flexible power stations and flexible consumers, CHP, storage and European electricity trading. And we are making it possible for these flexible capacities to be financed by the market'.

However, the minister added 'further to this, the White Paper sets out key elements for a capacity reserve. This is to provide additional "belt and braces" security for the electricity market 2.0.' So it will keep 4 GW of reserve capacity as a reliability guarantee. 2.7 GW of this will be old lignite coal plants, kept off-line unless needed and eventually closed entirely, when sufficient other capacity is available.

Taking these dirty plants out of normal use will help to reduce emissions, but critics worry that the new policy in effect offers a backdoor way to retain coal plants, at least for a while. It certainly does seem that coal will not be phased out quickly. The German Advisory Council on the Environment (SRU) says that 'an integrated energy policy should synchronise the phasing out of conventional power generation capacities and the increasing use of renewables', but warns that this will take time. However, it says that 'by 2040 at the latest, all coal-fired power stations should have been taken offline, preferably beginning with those with high specific greenhouse gas emissions'. It adds that by 2050, when renewables are planned to supply 80% of German electricity, 'the remaining power requirement must then be generated in flexibly controllable plants and can therefore not be provided by coal-fired power stations'.

This does seem to be quite a slow phase-out programme for coal, compared with the fast nuclear phase out, with eight nuclear plants already closed and the remaining nine plants to be shut by 2022. But the SRU says that 'if coal-fired power stations are taken offline at a faster rate this will lead at first to an increase in electricity prices' (SRU 2015)

That is a moot point. The spread of renewables has led to dramatic reductions in wholesale electricity prices, with, for example, low marginal cost PV solar squeezing higher cost fossil plants out of the peak demand market. This has been a problem for the fossil fuel companies, increasing their costs and reducing their profits, with some of this cost being passed on to consumers as higher retail charges. These charges have also been inflated by the rising costs of fossil fuels, especially imported gas, and although of late some of these costs have fallen globally (due to cuts in oil prices and US gas costs), the long-term global trend seems likely to be upwards, as fossil fuel reserves dwindle and higher emission charges are imposed. Although renewables may cost more initially, switching to their use should subsequently help decouple energy costs from the impacts of that trend. However, that may take time and meanwhile electricity prices have been a politically controversial issue in Germany, as they are in the UK and elsewhere. See box 6.1 for some views and data on electricity costs in Germany.

Although consumer price rises in Germany have only partly been due to green taxes and the like, as in the UK and as suggested in box 6.1, they may rise as renewables expand further. This is one reason why there has been pressure to slow the deployment of renewables. However, as we have seen, the counterview is that, in time, as the use of increasingly scarce and costly fossil fuel is reduced and renewable technology develops, generation costs should fall, squeezing fossil fuels out further. Slowing down renewables will just slow this process.

The SRU certainly see this process as an inevitable result of the transition to renewables: the market should gradually see off old base-load plants, since 'it would be uneconomical to continue to burn fossil fuels when power can be generated from technology operating without fuel costs.'

However, this transition may not in fact be rapid, automatic and without problems. For the moment, in the absence of high carbon prices (since, with weak carbon caps set, the EU Emission Trading System does not work well), coal is relatively cheap and electricity export markets are buoyant, so not all the coal plants may go. Indeed, it could mainly be the gas plants that go, squeezed out of the daytime peak market by marginal cost PV. That could be a problem, at least in the short to medium term, since they are cleaner and well suited to balancing.

It is not clear exactly how market pressures will keep some gas plants running as peaking plant back-up, while other, or additional, balancing systems, are developed. The SRU simply asserts that 'for new plants, the market itself will introduce the necessary structural change from base load power stations to peak load plants in the

> **Box 6.1. Electricity changes and costs in Germany.**
>
> While some say that electricity prices in Germany have risen because of the nuclear phase-out programme—which has allegedly led to more coal use—and the extra spending on renewables, the reality is more complex.
>
> Rapid change is certainly underway in Germany, with new technologies replacing the old ones. Renewables are now producing more electricity annually than has been lost from the nuclear closures, and more than from nuclear overall. It is true that some new coal plants have been built, with these being long-planned and more efficient than the ones they replaced, so carbon emissions have not suffered significantly; after an upward blip, they fell, as renewables expanded and demand fell. Although coal is still being burnt, the expansion of renewables is pushing the output from some coal plants out of the German market; it is claimed that the coal plants are increasingly serving the lucrative export market: German exports have risen to 6% or more of its electricity. Most 'greens' object to the continued use of coal and want the coal plants closed. Some old ones have been, but the EU Emission Trading Mechanism, which is meant to put a surcharge on coal use, has not proved effective as yet (Kunze and Lehmann 2015).
>
> Although most of the increase in electricity prices has been due to other factors, the charges imposed for deploying renewables have played a part. However, overall, the Agora Energiewende group claims that, after significant increases in previous years, household electricity prices in Germany have been relatively stable since 2013: the renewables policy surcharge (EEG) actually fell slightly in 2015 to 6.17 EUR cents kWh^{-1}, from a peak of 6.24 in 2014. Agora notes that renewable deployment lowers wholesale market prices, partly compensating for the increase in the EEG surcharge. Nevertheless, this is likely to rise again as renewables expand, and is expected to increase moderately until 2023, reaching about 7.20 EUR cents kWh^{-1}, before falling to 4.4 in 2035. Agora says that the current costs of supporting renewables include large historic development costs; future costs will be moderate. It puts the integration/back-up system cost for wind and solar at 5–20 EUR MWh^{-1}, which is small compared with their generation costs of 60–90 MW^{-1}, with the highest total still coming to less than that for new nuclear, put at 113 EUR MWh^{-1}, based on the UK Hinkley CfD strike price (Graichen 2015, Agora 2015a).

medium term' and looks to a future when 'rather than providing the base load, conventional power stations will be used to meet the residual load after electricity from renewable sources has been fed in' (SRU 2015).

Despite issues like this, with its major commitment to renewables (over 80 GW of wind and PV solar far and an overall 32% renewable contribution), Germany is obviously a test bed for policies that may have relevance across the world if they follow the German lead to a renewables-based future. As can be seen, there may be economic problems in the short term, and although this is important, it should not shape longer-term thinking. Indeed, in a welcome, and maybe brave, move away from short-term market thinking, the SRU says that 'Germany cannot and should not compete to offer the lowest electricity prices, but should rather focus on innovative, high-value products and processes that are environmentally compatible'.

Looking at the transition positively, it should build lucrative green markets to more than replace the jobs lost as fossil and nuclear are phased out, and avoid costly fossil fuel imports. And, of course, it should help to reduce the costs and impacts of climate change for everyone.

All that said, the proposed new electricity market 2.0 does represent something of a shift to the right politically, or at least to a more short-term market-orientated approach. The white paper says it is 'particularly important to make sure that pricing on the electricity market is not interfered with' and that it is 'making an explicit commitment to the liberalised European electricity market'.

That is clearly a free market approach, but the government claims that it will work and offers three justifications for its new national policy: 'Electricity market 2.0 firstly ensures security of supply, secondly is cheaper, and thirdly enables innovation and sustainability.'

We shall see. Can price signals and competition really hold it all together, and ensure that enough balancing capacity, including demand management, is available? The devil may be in the detail, e.g. there are some fudges and interventions on combined heat and power: the government will provide state aid to CHP plants whose economic viability is at risk, and promote the conversion from coal-fired to gas-fired CHP. Moreover, in all the talk about flexibility, energy saving gets short shrift. Will markets make it happen? And also reduce imports?

The new policy also sets a new context for the deployment of renewables. Whereas, at one time, guaranteed-price feed-in tariffs provided the main support mechanism, the Renewable Energy Sources Act, amended in 2014, now requires new installations to sell electricity from renewables directly to the market. The white paper says that at present approximately 70% of the electricity generated from renewables is sold directly to the market. By 2020, this share will rise to approximately 80%, according to current estimates. That will clearly help them fit into the proposed new market system, but there are many details yet to be sorted. Some projects, like offshore wind farms, are to be subject to contract auction processes. So it is hard to say how well they will fare.

One of the government's main concerns has, of course, been to reduce the cost to consumers. Slowing down and removing the feed-in tariffs will have done that, at least in the short term. Whether the new competitive market will achieve that long term, while still pushing renewables ahead, remains to be seen. Some fear that the move away from feed-in tariffs to contract auctions will not result in savings, but may slow effective deployment (Toke 2015).

6.3 Flexibility, base-load and market design

Leaving aside the nuances of electricity market economics, what does the German approach tell us about likely overall technology development patterns, and grid balancing in particular?

The SRU says that, with a large renewable contribution to energy supply and other balancing measures, 'in the near future, there will only be a need for flexible peak-load power stations which can contribute to the provision of a widely fluctuating residual load.' Along similar lines, a study by the Fraunhofer Institute for Agora Energiewende concluded that, in the medium term, 'base load capacities will decrease relative to those of today, while peak load and mid-merit capacities will increase'.

All of which means a very different approach to system design and management. As the Fraunhofer Institute put it: 'renewables, conventional generation, grids, the demand side and storage technologies must all become more responsive to provide flexibility' (Agora 2015b).

The flexibility concept certainly does provide a way through the often complex maze of energy technology choices. It suggests that base load is a problem, rather than, as it has been seen in the past, a vital necessity for ensuring grid security. As the SRU concluded, while 'in the transitional period, flexible gas-fired power stations will have a significant role to play', longer term 'there will be no need for base-load power stations, i.e. power stations which for technical or economic reasons should operate at a constant production level'.

That is not a view that is likely to be accepted in UK government circles, given the UK commitment to build a series of new large nuclear plants, 16 GW by around 2030, and perhaps much more later. Arguably, these plants will make it harder to balance the renewables component of the UK energy mix, which may reach 30 GW in the 2020s and more later, since, in order to recoup their large capital costs, the nuclear plants will have to be run 24/7. At present, there are no plans for them to load follow. So, unless extensive storage or export options are available, when demand is low (e.g. it can fall to around 20 GW at night in summer), at times, some renewables will have to be curtailed. Put simply, large inflexible base-load nuclear and variable renewables do not fit well together on the same grid.

While some therefore say that base-load is not the way ahead, limiting it and relying on variable sources does mean that balancing has to be upgraded. No single one of the balancing options we have looked at seems likely to be sufficient. There is always the risk of looking for a single 'silver bullet' solution. As the International Renewable Energy Agency put it in a report on energy storage, which, as we have seen, is sometimes depicted as the best answer, 'energy storage is only one of many options to increase system flexibility', including interconnectors, demand response, smart grid technologies and new pricing mechanisms (IRENA 2015).

As we have seen, they all have pluses and minuses. The task now is to select the right mix, as far as is possible, and then ensure that the necessary capacity is made available. Central intervention by government is no longer popular, with market-based systems being favoured. However, as we have seen, there are some disagreements about market design. In terms of providing support for balancing, it is not clear which of the two different versions being developed in the UK and Germany is the right approach.

In both cases, the costs are passed on to consumers. In the UK case, this is done directly, via a capacity market levy, while in the German case, it is done indirectly by the utilities, to defray the cost of providing balancing. It is unclear which will be the most effective approach and whether extra costs will have to be met. For example, there may be a need for further claw-backs of system costs. In the UK, in late 2015, in a speech on 'resetting the energy market', Amber Rudd, the Energy Secretary, said 'we want intermittent generators to be responsible for the pressures they add to the system when the wind does not blow or the Sun does not shine'. Clearly seeing this as part of a market approach, she added that the extension of responsibilities was necessary, since 'only when different technologies face their full costs can we achieve a more competitive market' (Rudd, 2015).

As we have seen, there are concerns that the UK capacity market approach will disrupt the wider energy market, while not delivering effective non-fossil balancing. The UK government, however, seems confident that this will not be a problem and that its capacity market system can deliver balancing services effectively (see box 6.2). Time will tell if they are right.

Whichever approach is adopted, the underlying aim is to provide flexibility as a way to respond to variable renewable energy (VRE) and that is clearly a good principle. As the World Bank noted in its review of the field, 'the flexibility of the overall power system will need to be increased as the share of VRE reaches higher levels'. It said that flexibility can be provided 'through additional interconnections to give systems operators access to a wider pool of demand and supply options; by implementing demand response measures to provide flexibility in demand; through optimizing and adding flexibility in supply, such as can be provided by NG-fired generation technologies; and/or through incorporating energy storage to act as additional demand through charging when there is excess energy, as well as additional supply through discharging when demand exceeds generation capacity, as needed' (Martinez and Hughes 2015).

It said that to help, 'the value of flexibility in the system should be recognized through policy and regulation, and remuneration mechanisms for flexible capacity should be defined'. Although it said that, 'for the most part, flexibility requirements should be technology agnostic in the absence of a strong reason to use a specific technology', it saw a 'close link between scale-up of VRE and natural gas-fired power generation in many countries'.

As we have seen, that may only provide an interim and partial solution. What is needed is the development of the other more advanced balancing options, and their integration with new and existing renewables into a fully sustainable and reliable energy supply and utilisation system. This will be quite a challenge.

6.4 Making the change globally

This chapter has looked at how support for grid balancing systems is being promoted in the EU and at some views on the relative merits of the various options and support policies. The same debate is going on elsewhere.

As noted earlier, some ambitious scenarios have emerged in the USA (Jacobson *et al* 2015). The scene was in effect set by the renewable electricity futures report produced by US National Renewable Energy Laboratory in 2012, which said that: 'renewable electricity generation from technologies that are commercially available today, in combination with a more flexible electric system, is more than adequate to supply 80% of total US electricity generation in 2050 while meeting electricity demand on an hourly basis in every region of the United States' (Mai *et al* 2012).

> **Box 6.2. The role of markets in balancing: the UK government view.**
>
> The UK government's response to an EU consultation on energy market design said that 'in theory, the electricity market should provide incentives for investment in sufficient reliable capacity. However, the GB electricity market—like many others—faces market failures, exacerbated by new pressures such as the increase in intermittent forms of energy, which mean there is a significant risk that this will not be the case', and it identified market failure problems in relation to generation adequacy, 'that scarcity pricing alone may not fully be able to address'.
>
> By contrast, it said that the GB capacity market could and 'will not distort the functioning of the GB electricity market. Indeed, the Capacity Market has been designed to complement the market signals that already exist aimed at ensuring capacity adequacy in a market that allows for efficient dispatch decisions.' Its competitive auction approach also minimised the cost to consumers.
>
> So the UK government implicitly rejected the German approach; the UK system was successful, involved a range of players, including independent generators, and could be expanded to support a wider range of balancing options. Consequently, the UK government did not see 'significant obstacles to fully integrating renewable energy generators into the market. In the UK, renewables (apart from installations smaller than 5 MW that are supported through the small-scale feed-in tariff scheme) already face full balancing responsibilities.' The capacity market would buttress that. Evidently confident in the merits of this approach, the report even suggested that it might be possible for other EU players to join in with projects at some stage (DECC 2015a).
>
> Interestingly, the second round of the UK capacity market auction process at the end of 2015 led to much the same outcome as the first round (which was for 2018–19). Once again, most of the 46 GW of contracts in the new round (for 2019–20) went to old gas, coal and nuclear plants (over 36 GW in all), with only 475 MW to demand-side balancing, and nothing to new interconnectors, although the existing 1.8 GW link was re-contracted (DECC 2015b). Surprisingly, only a few new gas plants managed to get contracts, just 810 MW, despite the government being keen to expand gas. 4.6 GW of new gas project bids were evidently turned down. It could be that the 18 UK pound MWh^{-1} subsidy being offered by the capacity market in this round (£1.40 less than in the first round), while still a helpful windfall for existing plants, was not enough to support many new plants. One result seems to be that around 870 MW of cheap but dirty diesel plants were contracted to partly fill the gap, something that did not go down well with environmentalists (Sandbag 2015).
>
> A House of Lords Select Committee review had called for adjustment to the capacity market, so as to better stimulate the uptake of new balancing options such as demand side response and storage, but in its response the government said, 'the Capacity Market auction is technology neutral. The auction mechanism is driven by cost-effectiveness to determine the most suitable providers and the introduction of specific targets in the Capacity Market would lead to less competitive auctions increasing costs to consumers' (DECC 2015c).

There is a long way to go before anything like that is realised (renewables only supply around 15% of US electricity at present), but progress has been made on some aspects of the flexible balancing that would be needed. For example, some large smart grid assessment projects have been set up, with 3.7 billion US dollars of the post-economic crash 'American Recovery' funding going to 100 projects (DoE 2013). Some early work, like that carried out at the University California, Berkeley and elsewhere, has looked at the demand-side impacts of variable pricing in California, though there have been problems with dealing with 'additionality' issues, i.e. would some of the effects seen have happened anyway, without the extra price signals (Davis 2013)?

On the supply side, there have been many studies of the integration issue, and a substantial body of research work now exists (Apt and Jaramillo 2014), along with much practical experience of grid balancing and linked regulatory developments (Jones *et al* 2014). Curtailment issues have been a key concern, and in some cases they (and balancing issues generally) have been addressed by state or local-level grid enhancement projects, or by adjustments to scheduling and reserve policies. For example, the Electric Reliability Council of Texas (ERCOT), the independent system operator that manages the electric power grid serving 85% of the load in the state of Texas, has developed reserve management practices that enable it to deal with the state's large renewable capacity (Dumas and Maggio 2014). On an even larger scale, the PJM regional transmission organisation has developed experience with forward capacity market planning and demand response covering more than 13 states (Ott 2014).

The spread of renewables and the development of new balancing technologies, including new local and domestic storage options, is leading to more detailed assessment of supply and balancing issues, in the search for optimal mixes. While some see distributed storage as a 'game changer', as was noted in chapter 5, in one study, national supergrid 'electron superhighways' were seen as the least costly integration option, likely to enable extensive renewable development without the need for much storage (MacDonald *et al* 2016).

The debate continues, but in time these studies may lead to new support mechanisms, of the sort being tested in the EU, although the US situation differs from that in the EU. Given the generally larger distances, long-distance cross-country grid linking is less common, and even regional/area links can be weak. For the moment, national supergrids may still be some way off, although attempts are being made to improve the grid system, with, for example, a three year 220 million US dollar US-wide grid modernisation programme being announced in 2016. There are also some large one-off long-distance projects. For example, as was noted earlier, there is an ambitious proposal for an HVDC project linking wind generation in Wyoming to cavern compressed air storage in Utah and electric power delivery in Los Angeles, in a grid loop of over 1000 miles in all.

Distances are, if anything, even greater in China, and that has proved to be a problem for integrating wind energy into the national system, with supergrids being one possible solution, despite the high cost (see box 6.3).

Problems like this are being addressed by an interesting collaborative project with the USA on flexible grid and energy system development, which draws on US expertise and experience and aims to support China's renewable programme (Milligan *et al* 2015). The scale of expansion that is thought to be possible is certainly significant. The Chinese National Renewable Energy Centre, which is

> **Box 6.3. Wind curtailment in China.**
>
> The expansion of wind power in China initially occurred in something of a chaotic burst in response to capacity targets and state imperatives, and ran into trouble since insufficient attention had been played to providing grid links, which had mostly been left to poorly resourced local and regional government agencies to deal with. The result was that, although very significant wind capacity was installed, reaching over 42 GW by the start of 2011, only about 31 GW was grid-linked. Many of these projects—most of which were in remote areas in the north west that were poorly served by grid links— were often unable to dispatch their full potential output to users, most of whom were in the major urban areas on the south east coast (Andrews-Speed 2015).
>
> The problems have worsened as more wind capacity has been installed. China now has over 145 GW. Around 12% of total wind output was curtailed in 2014, with this rising to 15% in the first half of 2015, and the figures are even larger for the wind projects in the remoter locations, far from the major urban energy demand centres. In the first half of 2015, curtailment exceeded 43% in the province of Jilin, 31% in Gansu, 28.8% in Xinjiang and 20% in West Inner Mongolia (Hurlbut *et al* 2015).
>
> Central government policy changes have sought to alleviate this problem. Building HVDC links is one obvious answer, as has already been done to link in large hydro, although wind projects are much more diffuse. So there is a need for local grid upgrades as well as system management improvements.

involved with the project, has suggested that China could get 85% of its electricity and 60% of all energy from renewables by 2050 (Darby 2015). Its detailed roadmap looks in particular at integration issues, stressing the need for flexible supply- and demand-side systems and inter-province grid connections (CNREC 2015).

Africa also has major renewable potential but, as in parts of China, grids are very undeveloped, and off-grid diesel generation is common, although increasingly renewables are being deployed locally. As noted in chapter 4, moving beyond off-grid systems may be hard in remote areas, and local micro grids may make the most sense in the immediate future, backed up by local storage, although there are plans for supergrid corridors across the continent. However, there are many political and logistical issues, as well as significant funding concerns (IRENA 2014).

The ambitious Desertec initiative to link up concentrated solar power projects in North Africa with the EU via HVDC supergrids was mentioned earlier. As was noted, schemes like that open up many issues, and its fate (it has been sidelined for the moment) indicates how difficult it is to obtain agreement on major new projects; support from Desertec's backers, mostly in Germany, has dwindled (ScidevNet 2013).

This may not have surprised supergrid pioneer Dr Gregor Czich from Kassel University, who has always argued that, for the present, wind energy technology is much more relevant than the still very costly concentrated solar power systems, while it offers an equally large untapped resource (Czich 2011). For example, he put the wind resource in Morocco, west of the Atlas mountains, at 120 GW, and that in Mauritania at 105 GW. Further north, he identified a 350 GW wind resource in northern Russia and Siberia, and to the east, a 210 GW wind resource in

Kazakhstan. A recent study even suggested that if these huge resources were linked up by a supergrid system, they could supply Russia and the neighbouring central Asian countries with all the electricity they would need by 2030, at competitive costs (Bogdanov and Breyer 2015). The EU has its own vast wind resources, onshore and offshore, but in time we may see supergrids stretch to integrate some or all of these large, more remote wind resources beyond it borders. However, large-scale concentrated solar power costs are falling, especially for large-scale focused PV solar (CPV), so remote, centralised solar collection may, in time, also be linked in.

While each of these options has integration and balancing implications, all may be followed up, although too much of an emphasis on giant generation schemes and long-distance transmission is probably not appropriate at this stage. More rapid progress can perhaps be made incrementally, in many parts of the world, with small- and medium-scale projects, mini grids and local storage systems, as IRENA has argued for in the case of Africa. However, this does not mean that the focus should just be on local individual projects: these must be set in a wider context. IRENA says that smart grids can play an important role, 'facilitating smooth integration of high shares of variable renewables; supporting the decentralised production of power; creating new business models through enhanced information flows, consumer engagement and improved system control; and providing flexibility on the demand side'. So it concludes that 'policies and regulations need to be developed for smart grids and renewable energy sources as soon as, if not before, large-scale deployment takes off' (IRENA 2013).

6.5 Institutional challenges

As can be seen, although there is a range of technical possibilities for renewable supply and flexible grid balancing around the world, there are also strategic issues concerning which should be used, and how and when they should be deployed. Some issues are purely technological, but many of them may also involve and require institutional and policy changes.

An NREL study (produced as part of the collaborative programme with China mentioned above) noted that, in relation to grid balancing, technical and institutional changes may both be important. Moreover, they often interact, sometimes negatively: 'physical flexibility can be hampered by something as simple as the market settlement process' (Milligan 2015). The report provided a list of possible balancing measures, and indicated which they felt would require physical/technical changes, which they thought would need institutional changes, and which they believed would need both (see table 6.1).

Some of the changes in approach may be quite hard, given local conditions, with some potential technical and policy conflicts emerging. For example, another report in the NREL–China series notes that, in China, administrative priority is 'given to the operation of combined heat and power (CHP) units so that heat supply to local residents can be guaranteed during the long winter months'. As a result, 'thermal units often displace wind generation. Additionally, the winter demand for heat reduces the ability of CHP units to ramp down, making variable generation (VG) integration more difficult' (Hurlbut *et al* 2015).

Table 6.1. Flexibility measure development. Physical or institutional? Data taken from Milligan *et al* (2015).

Larger balancing areas	Both
Access to neighboring markets	Both
Faster energy markets	Institutional
Regional transmission planning for economics and reliability	Both
Robust electrical grid	Physical
Improved market design	Institutional
Demand response	Both
Geographically dispersed VG	Physical
Strategic VG curtailment	Both
VG forecasting effectively integrated into operations	Both
New flexibility for ancillary services/products	Institutional
Sufficient reserves for VG event response	Physical
Flexible conventional generators	Physical
Primary frequency response, inertial response and response to dispatch signals with new VG technologies	Both
Storage	Physical

As we have seen, CHP is valuable in itself, as a low carbon local heat source, but also as a way to balance power and heat use against changes in demand and renewable supply. Similar conflicts have emerged in the EU, in Denmark and Germany especially, although, in that case, CHP/district heating projects are being squeezed out of the energy market by cheaper wind power. Support measures may have to be adopted to limit that. Clearly, establishing and then managing the right mix of systems will be hard.

Deciding who pays for them, and the full balancing costs, may be even harder. For example, as we have seen, the deployment of marginal cost variable renewables disrupts the market for existing energy technology, creating a new market profile, and it may undermine the profits of the utilities. Should they seek compensation? Some of this loss is just due to fair competition, but some is a cost that will have wider implications, due to changes in the scale and utilization of back-up capacity that may be needed for balancing. As we have seen, this is being addressed to some extent by the capacity market in the UK and, hopefully, the wider market in Germany. So in the end, either way, consumers will pay. But should they pay for the changes in the so-called profile costs? Box 6.4 examines the costs of change.

In addition to these financial support and market cost issues, there are also public policy challenges associated with the various balancing options. For example, the spread of smart grid/demand management systems will open up information access and privacy issues. The institutional and policy challenges may be harder still when the focus is more international, for example in relation to cross-border energy transmission and balancing. This raises a range of policy issues, as Scholten and Bosman point out in their interesting exploration of the geopolitics of renewables: 'How to manage the intermittency of power generation in cross-border networks; how will

> **Box 6.4. The costs of change.**
>
> As we saw in chapter 5, Agora Enegiewende claimed that the profile change costs resulting from the reduced utilization of conventional plants amount to a maximum of 13 EUR MWh^{-1} with a 50% share of wind and solar energy in Germany. As we also saw, others put it higher. Who should pay for that? Agora noted that 'experts disagree on whether the "utilization effect" can (and should) be considered as integration costs'. They say that 'adding any type of new power plant reduces the utilization of existing power plants. It has been debated whether this effect can (and should) be considered as an integration cost and how the value of power plants and/or lost revenues of operators can be quantified. At higher penetration rates, the effect from new wind and solar power plants may differ significantly from those of new base-load power plants. The former requires more dispatchable capacity in the system and a changed pattern of residual demand, leading to a shift of power production from base load to mid-merit and peak load power plants'.
>
> Each option presents utilities with very different costs. Agora comments that 'quantifying the cost of these effects depends largely on the perspective taken, on the system considered and on the definition of costs applied'. It suggests a total system costs approach, similar to the system LCOE proposed by the Potsdam Institute, combining generation costs and integration/balancing/system costs. They say that this approach 'can assess the total costs while avoiding the controversial attribution of system effects to specific technologies' (Agora 2015c).
>
> Certainly this approach does provide a standard measure, but it does not avoid potential disputes over the make-up of the costs and who pays for the various elements, including the changed market/supply profile costs. The complexity will not disappear when the sub-costs are combined in this way. Indeed, it could become even more complex. For example, it is possible that in future choices will have to be made between variable renewables, each with effectively zero marginal generating cost. The choice may then depend on the balancing and system costs, which may differ if the sources have different weather-defined characteristics, or on the environmental costs, which again may differ slightly. The total system cost approach may have to be enlarged even further, assuming that a market-based assessment approach is retained.

damages in one area incurred by fluctuating power in another area be resolved; what new modes of operating these systems may be required?' (Scholten and Bosman 2016).

A recent EU e-Highway2050 transmission study identified some of the specific economic management issues within and across the EU, arguing that that there is an urgent need 'to complete the internal energy market and to ensure regional market integration in all time-frames, forward, day-ahead, intra-day and real-time', with 'well-designed balancing markets' being a key requirement (e-Highway2050 2015).

Clearly, if such systems are to develop on a significant scale, there will be a need for careful planning and integration, as well as political negotiation. Indeed, there may be a need for central political intervention. For example, some of the large energy utilities may not be too enthusiastic about supergrids, since widespread cross-border trading may undermine their regional market control, so, within the European context, it may fall to the European Commission to take a lead, if more rapid progress is to be made. Given that major supergrid expansion would also involve geopolitical issues, as well as many economic management issues, that is probably inevitable anyway. As the EU e-Highway study noted, not only will it be crucial 'to incentivise market actors to ensure correct and rational behaviour in order to tackle ever-increasing system security aspects', there will also be a need for 'further regional security monitoring and control mechanisms closer to real-time over larger geographical areas' (e-Highway2050 2015).

It is likely to be similar elsewhere: coherent direction and planning will be needed. For example, as noted earlier (box 6.3), the development of wind energy in China has met with problems due to the lack of grid integration, arguably highlighting the need for more coherent planning and management.

Some of the problems facing renewable development may be more widely political. Although not unique in this respect, China also provides an example of the need for institutional policy change, with wide political implications. The wind development problems mentioned above might be taken to indicate a failure of China's bureaucratic central planning system, in which case, while some might say that *more detailed* planning is needed, others may look to markets to provide better incentives.

Certainly, looking at the overall management of China's energy system, the US National Renewable Energy Lab report, produced as part of the collaborative programme with China mentioned earlier, noted that 'China does not have an organized, actively traded electricity market. The annual amount of inter-provincial power transmission is determined administratively each year by the local governments involved. Fuel prices are also administered and change very slowly [...] Unit commitment is not done on the basis of economic merit. Rather, schedules are set yearly, monthly, and weekly in a hierarchical manner from national, regional, and provincial to county and municipal levels. Traditionally, managing generator schedules focuses on making steady progress towards yearly contracts and administratively allocating generation hours among several power plants' (Hurlbut *et al* 2015).

However, measures promulgated in 2007 started China on a path towards an 'energy saving dispatch' approach, and NREL say that 'in early 2015, China embarked on a new round of power sector reform, with goals to increase the use of market mechanisms for ancillary services, direct electricity sales, and demand-side management programs'. It may take a while to introduce the sort of market competition the USA has, and not everyone will see that as ideal. Arguably, markets are not always the solution and some *more effective* planning, central or otherwise, might not be amiss, and this may also be the case in the US and elsewhere.

That is clearly a matter of political preference. However, it could be argued that, given ongoing technological change and a shift away from central utilities, a more decentralized planning and management system is needed. Indeed, some have called for more decentralised policy making and regulation, that is open to a wider range of participants (Mitchell *et al* 2015).

Arguably, most of what is needed involves technical and institutional change, but the policy framework may also need to change. Certainly, in addition to specific market design issues, as renewables expand around the world and new grid and balancing systems emerge to cope with that, there is likely to be a political dimension to at least some parts of the process of developing, planning and deploying the new system.

6.6 The challenges of change

In a 2011 report on the challenges of intermittency in north west European power markets, based on an 2009 study of the impacts of wind power in the UK and Ireland, the Poyry consultancy warned that as the use of variable renewables expanded, 'wholesale prices are likely to become increasingly volatile, risk to investors in thermal plant could increase dramatically and operational regimes of power stations like gas-fired CCGTs will become erratic as well as their load factors reducing' (Poyry 2011).

They have been proved right. However, it is also true that measures can be taken to respond to some of these problems and to integrate renewables into the power system. Indeed, if that is not done, then the situation could become even worse. As a US National Renewable Energy Lab report put it, if flexibility needs are not met, the system may experience reliability and economic consequences, including 'dropped load, VG curtailment, deviations from the schedule of area power balance, frequency and voltage excursions due to over- or under-generation, negative market prices, and price volatility' (Milligan *et al* 2015).

Fortunately, it seems clear that a range of grid balancing methods exist that can limit these problems. They can be further developed and deployed around the world as renewables expand. They will mostly be pioneered in the industrialised countries, but will soon become relevant elsewhere. Not all the lessons learnt from the pioneering countries may be relevant, since contexts vary dramatically. That may be especially true of approaches to supporting the introduction of balancing systems: they need to be tailored to local conditions, opportunities and constraints. For example, market-based approaches may not be the best starting points for countries with undeveloped energy markets, as in much of Africa.

In terms of the technology, as this book has shown, while most observers are confident that fossil back-up will remain a mainstay as renewables expand, buttressed by pumped hydro where available, there are still questions as to whether the other newer systems will prove effective at balancing grids, technically and economically, in all contexts.

Technically, they all could be, to varying degrees, but economically, it is less clear. To some extent, that is because current economic frameworks reflect a narrow interpretation of social and environmental value. Progress is nevertheless being made around the world, with balancing issues being increasingly well researched and understood, as even a cursory glance at some of the reports and studies that have emerged will indicate (Jones *et al* 2014, Apt and Jaramillo 2014). As we have seen, there are some inspiring projects underway globally, and there are ambitious proposals for more of these. There are also some positive visions for the future. For example, Sorensen outlined some radical '100% renewable' scenarios, with balancing issues to the fore, for China, Japan and South Korea (Sorensen 2014). Similar ones, with high renewable targets, have emerged for other countries, including India (WWF 2013).

Inevitably, as attempts are made to expand and balance renewables, there will be problems. In some contexts, there may be conflicts between the balancing options. Certainly, there will be winners and losers. For example, dynamic demand management and storage may challenge the supply market in the same way that prosumer PV has. Supergrid power exchanges would also challenge existing market relations. However, that is what you get with change—challenges to the status quo.

The final chapter looks further at the challenges. They need to be recognized. But what matters now is to face them. Technological solutions are not the only answer, but they can help. Moreover, far from being new and disturbing, some of the ideas now being put into practice have a long history. For example, in a talk given in Cambridge in 1923, J B S Haldane predicted that 'The country will be covered with rows of metallic windmills working electric motors which in their turn supply current at a very high voltage to great electric mains. At suitable distances, there will be great power stations where during windy weather the surplus power will be used for the electrolytic decomposition of water into oxygen and hydrogen' (Haldane, 1923).

Over 90 years later, we have a UK company, ITM Power, installing systems like that (power-to-gas hydrogen PEM cells) in Germany (ITM 2015). Whatever next?

Chapter summary

1. Various approaches to ensuring that balancing services are available are being used around the world.
2. There are disagreements about which is best, based on views about the role and efficacy of market competition.
3. Optimal approaches are likely to require the consideration of total system costs rather than just minimization of the short-term costs of individual supply and balancing components.
4. Making the changes needed may require difficult institutional changes.

References

Agora 2015a *Understanding the Energiewende* (Berlin: Agora Energiewende) www.agora-energiewende.de/fileadmin/Projekte/2015/Understanding_the_EW/Agora_Understanding_the_Energiewende.pdf

Agora 2015b The European power system in 2030—flexibility challenges and integration benefits *Fraunhofer Institute Report for Agora Energiewende* www.agora-energiewende.org/service/publikationen/publikation/pub-action/show/pub-title/the-european-power-system-in-2030-flexibility-challenges-and-integration-benefits/

Agora 2015c *The Integration Costs of Wind and Solar Power* (Berlin: Agora Energiewende) www.agora-energiewende.de/fileadmin/Projekte/2014/integrationskosten-wind-pv/Agora_Integration_Cost_Wind_PV_web.pdf

Andrews-Speed P 2015 *Energy Governance in China: Transition to a Low-Carbon Economy* (Basingstoke: Palgrave)

Apt J and Jaramillo P (ed) 2014 *Variable Renewable Energy and the Electricity Grid* (London: Routledge)

BMWi 2015 An electricity market for Germany's energy transition *White Paper* (Berlin: Federal Ministry for Economic Affairs and Energy) www.bmwi.de/EN/Service/publications, did=721538.html

Bogdnov D and Breyer C 2015 Eurasian super grid for 100% renewable energy power supply: generation and storage technologies in the cost optimal mix *ISES Solar World Congress 2015 (Daegu)* www.researchgate.net/publication/283713531_Eurasian_Super_Grid_for_100_Renewable_Energy_power_supply_Generation_and_storage_technologies_in_the_cost_optimal_mix

CNERC 2015 China 2050 high renewable energy penetration scenario and rodamap study *China National Renewable Energy Center* www.rff.org/files/sharepoint/Documents/Events/150420-Zhongying-ChinaEnergyRoadmap-Slides.pdf

Czisch G 2011 *Scenarios for a Future Electricity Supply: Cost-Optimized Variations on Supplying Europe and Its Neighbours with Electricity from Renewable Energies* (London: IET) www.theiet.org/resources/books/pow-en/scenarios.cfm

Darby M 2015 China's electricity could go 85% renewable by 2050 *Climate Home* (24 April 2015) www.climatechangenews.com/2015/04/22/chinas-electricity-could-go-85-renewable-by-2050-study

Davis L 2013 20/20 vision *University of California, Berkeley, Haas Research* Paper Overview http://energyathaas.wordpress.com/2013/10/21/2020-vision/

DECC 2015a UK government response to the Energy Market Design consultation (London: Department of Energy and Climate Change) www.gov.uk/government/publications/uks-response-to-european-commission-consultation-on-energy-market-design

DECC 2015b *Securing Future Electricity Supplies* (London: Department of Energy and Climate Change) www.gov.uk/government/news/securing-future-electricity-supply

DECC 2015c Government response to the House of Lords Science and Technology Select Committee inquiry: the resilience of the electricity system (London: Department of Energy and Climate Change) www.gov.uk/government/publications/government-response-to-the-house-of-lords-science-and-technology-select-committee-inquirythe-resilience-of-the-electricity-system

DoE 2013 *Insight on Smart Grid Customer Engagement* (US Department of Energy) www.smartgrid.gov/sites/default/files/VoicesofExperience_Brochure_9.26.2013.pdf

Dumas J and Maggio D 2014 Electric reliability council of Texas case study: reserve management for integrating renewable generation in electricity markets *Renewable Energy Integration* ed L Jones (London: Elsevier)

Ecogrid 2015 *Ecogrid* www.eu-ecogrid.net/images/News/131004_%20edk%20a4_ecogrid%20eu%20project_web.pdf

e-Highway 2050 2015 *Europe's future secure and sustainable electricity infrastructure: e-Highway2050 project results (RTE, Paris)* www.e-highway2050.eu/results

Graichen P 2015 Insights from Germany's Energiewende *Agora Energiewende Presentation* www.agora-energiewende.de/fileadmin/Projekte/2015/Understanding_the_EW/Key_Insights_Energy_Transition_EN_Stand_7.10.2015_web.pdf

Green Power Island 2015 *Green Power Island* www.greenpowerisland.dk

Haldane J 1923 Lecture to the Heretics Society, University of Cambridge (4 February 1923) (Later published as a book, *Daedalus; or, Science and the Future*)

Harrison L 2015 The trouble with capacity markets *Windpower Monthly* (1 November 2015) www.windpowermonthly.com/article/1218191/trouble-capacity-markets

Hurlbut D, Zhou E, Porter K and Arent D 2015 *Renewables-Friendly Grid Development Strategies: Experience in the United States, Potential Lessons for China* (Golden, CO: National Renewable Energy Labs) www.nrel.gov/docs/fy16osti/64940.pdf

IRENA 2013 *Smart Grids and Renewables* (Abu Dhabi: International Renewable Energy Agency) www.irena.org/menu/index.aspx?mnu=Subcat&PriMenuID=36&CatID=141&SubcatID=362

IRENA 2014 *Africa Clean Energy Corridor: Analysis of Infrastructure for Renewable Power in Southern Africa* (Abu Dhabi: International Renewable Energy Agency) www.irena.org/DocumentDownloads/Publications/ACEC%20Document%20V19-For%20Web%20Viewing-Small.pdf

IRENA 2015 *Renewables and Electricity Storage* (Abu Dhabi: International Renewable Energy Agency) www.irena.org/DocumentDownloads/Publications/IRENA_REmap_Electricity_Storage_2015.pdf

ITM 2015 Thüga power-to-gas plant *ITM Power* www.itm-power.com/project/thuga-power-to-gas

Jacobson M *et al* 2015 100% clean and renewable wind, water, sunlight (WWS) all-sector energy roadmaps for the 50 United States *Energy Environ. Sci.* **8** 2093–117

Jones L (ed) 2014 *Renewable Energy Integration* (London: Elsevier)

Kunze C and Lehmann P 2015 The myth of the dark side of Germany's Energiewende *REnew Economy* (18 February 2015) http://reneweconomy.com.au/2015/the-myth-of-the-dark-side-of-germanys-energiewende-94542

MacDonald A, Clack C, Alexander A, Dunbar A, Wilczak J and Xie Y 2016 Future cost-competitive electricity systems and their impact on US CO_2 emissions *Nat. Clim. Change* doi:10.1038/nclimate2921

Mai T, Sandor D, Wiser R and Schneider T 2012 Renewable electricity futures study: executive summary *US National Renewable Energy Laboratory Report* NREL/TP-6A20-52409-ES www.nrel.gov/docs/fy13osti/52409-ES.pdf

Martinez S and Hughes W 2015 Bringing variable renewable energy up to scale: options for grid integration using natural gas and energy storage *World Bank Report* http://documents.worldbank.org/curated/en/2015/02/24141471/bringing-variable-renewable-energy-up-scale-options-grid-integration-using-natural-gas-energy-storage

Milligan M, Frew B, Zhou E and Arent D 2015 *Advancing System Flexibility for High Penetration Renewable Integration* (Golden, CO: National Renewable Energy Laboratory) www.nrel.gov/docs/fy16osti/64864.pdf

Mitchell C, Woodman B, Kuzemko C and Hoggett R 2015 *Working paper: Public Value Energy Governance* (University of Exeter Energy Policy Group, 20 March 2015) http://projects.exeter.ac.uk/igov/working-paper-public-value-energy-governance/1

National Grid 2015 Capacity market auction, provisional results *National Grid* www.gov.uk/government/uploads/system/uploads/attachment_data/file/389832/Provisional_Results_Report-Ammendment.pdf

Oroschakoff K 2015 German winds make Central Europe shiver *Politico* (3 August 2015) www.politico.eu/article/strong-winds-in-germany-a-problem-in-central-europe

Ott A 2014 Case study: demand-response and alternative technologies in electricity markets *Renewable Energy Integration* ed L Jones (London: Elsevier)

PB 2009 *Powering the Future* (Newcastle upon Tyne: Parsons Brinckerhoff) www.pbworld.com/pdfs/regional/uk_europe/pb_ptf_summary_report.pdf

Poyry 2011 *The Challenges of Intermittency in North West European Power Markets* (Oxford: Poyry) www.poyry.com/news-events/news/groundbreaking-study-impact-wind-and-solar-generation-electricity-markets-north

Rudd A 2015 *A New Direction for UK Energy Policy* (speech) (London: Department of Energy and Climate Change) www.gov.uk/government/speeches/amber-rudds-speech-on-a-new-direction-for-uk-energy-policy

Sandbag 2015 UK Capacity mechanism proves an expensive flop that is fuelling investment in dirty power *Sandbag NGO* blog (11 December 2015) https://sandbag.org.uk/blog/2015/dec/11/uk-capacity-mechanism-proves-expensive-flop-fuelli/

Scholten D and Bosman R 2016 The geopolitics of renewables; exploring the political implications of renewable energy systems *Technol. Forecast. Soc. Change* **103** 273–83

Scidev.net 2013 Global desert energy project hit by key partner's exit *Scidev.net* (5 July 2013) www.scidev.net/global/desert-science/news/global-desert-energy-project-hit-by-key-partner-s-exit.html

Sorensen B 2014 *Energy Intermittency* (London: Routledge)

SRU 2015 *The Future of Coal through 2040* (Berlin: German Advisory Council on the Environment) www.umweltrat.de/SharedDocs/Downloads/EN/05_Comments/2012_2016/2015_09_KzU_14_Future_of_Coal.html

Toke D 2015 Renewable energy auctions and tenders: how good are they? *Int. J. Sust. Plan. Manage.* **8** 43–56

WWF 2013 *100% Renewable Energy by 2050 for India* (New Delhi: World Wildlife Fund for Nature/TERI) www.wwfindia.org/?10261/100-Renewable-Energy-by-2050-for-India

Chapter 7

Conclusion: all change

This short final chapter summarises and reprises some of the key issues and questions raised in this book. Can the change over to renewables be made without massive additional cost for balancing the variability that some exhibit? Can fossil fuel use be entirely avoided? What about nuclear power– can that play a role? It looks at some of the uncertainties and challenges ahead, and offers some tentative positive conclusions: the change is needed, and a shift to a reliable and sustainable energy future, including full grid balancing, is possible and should not be unduly disruptive or costly.

7.1 The balancing challenge

We are facing a need to change the way we generate and use energy due to the major social and environmental problems associated with the existing system. The use of renewable energy is widely seen as part of the answer. However, to some, the challenges ahead in trying to shift to renewable energy look daunting, especially in relation to dealing with the variability of some of the sources. Given this situation, it is understandable that some say renewables are not a reliable energy source, or can only be used by adopting prohibitively expensive balancing measures. For example, Heinberg relays the view that 'the costs of enabling solar and wind to act like fossil fuels are so great as to virtually cancel out these renewables' very real benefits' (Heinberg 2015a).

As this book has shown, this may not be the case. There are ways to deal with this variability and they may not be costly. Indeed, some of the balancing systems described may actually reduce the overall system cost. Smart grid demand management can avoid waste and better match supply to demand. Supergrid export of surplus electricity can earn billions. Moreover, the generation costs of renewables should fall and improved system flexibility should enable them to expand dramatically at low overall balancing cost. The Imperial College/NERA study we looked at in chapter 5 noted that in the maximal case of their UK model, 'where a high level of flexibility is available, we observe a massive shift in the generation mix towards renewable technologies, with about 90–95 GW of wind and PV capacity, reflecting the reduced integration cost of RES technologies enabled by enhanced flexibility' (Imperial/NERA 2015). Similar conclusions were reached in many of the other studies we have mentioned: flexibility can enable rapid, low-cost expansion, whether in the east or the west (Bogdanov and Breyer 2015, NOAA 2016).

However, as we have seen, not everyone agrees with this view. There are debates about how the cost of balancing renewables should be calculated, and so estimates of their scale can differ. Agora Energiwende noted that 'some calculations of this integration cost component yield very high results while other yield very low—or even negative—results, even when analyzing the same system and situation'. Their analysis, and that of the International Renewable Energy Agency (IRENA) and the International Energy Agency (IEA), favoured the latter view, whereas the Nuclear Energy Agency and the Potsdam Institute tended to the former view, which is also usually shared by the utilities: they see their costs rising. The methodological debate can become quite obscure, and certainly complex, in part due to different assumptions about the impacts on the residual conventional plants and how they should be costed and shared (Agora 2015a).

There is also the issue of recognising the large environmental, health and climate costs of using fossil fuel. If those costs are included, then the significance of the balancing costs reduces, with IRENA commenting that 'When externalities are taken into account, renewables are virtually always the cheapest option for society' (IRENA 2015). However, even if these costs are excluded, IRENA still sees onshore wind as having lower total costs, including full balancing and extra (spatial) transmission costs, than conventional or nuclear sources (figure 7.1).

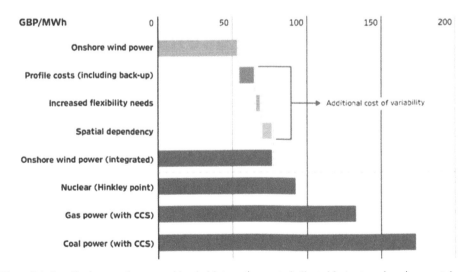

Figure 7.1. Levelised costs of energy with wind integration costs indicated but external environmental costs ignored. Reproduced with permission from IRENA (2015).

Moreover, quite apart from the likelihood that fossil and nuclear costs will rise, it seems (fairly) clear that, although making the transition to renewables may add some initial extra supply and balancing costs, these should fall as the renewable supply and associated balancing technology improves, and as new flexible systems, rather than old plants, take the strain. That said, the new system will look very different and to some extent it is unhelpful to speculate on exactly what the mix will be.

However, although it is not easy to summarise the complex interactions discussed in this book graphically, figure 7.2 (from iGov at Exeter University) presents an overview of the options (flexible fossil plants, storage, demand-side response, supergrid interconnectors, etc) and how they might interact in a scenario in which renewables supply the bulk of electricity.

Although figure 7.2 only represents, in simplified form, one possible arrangement, in effect it highlights many of the issues that have been discussed in this book. It assumes that renewables can usually meet average demand, and that on sunny and windy days they can meet maximum demand and much more, with the surplus being either exported or stored for use on days with no wind and no sun. It has demand-side response (deferring peaks) and imports being used to cope with peaks in demand and, along with the other balancing measures, helping to meet demand when renewable supply remains low, backed up by the use of gas to deal with demand peaks. However, the size of each balancing block is debatable, as is the residual role of fossil gas in the system.

Clearly, this depiction is very general. The final mix will vary with location and will depend on business and political decisions that are yet to be made. We can, however, risk some general conclusions about possible mixes and the technical viability of the balancing options, and look at some of the potential problems.

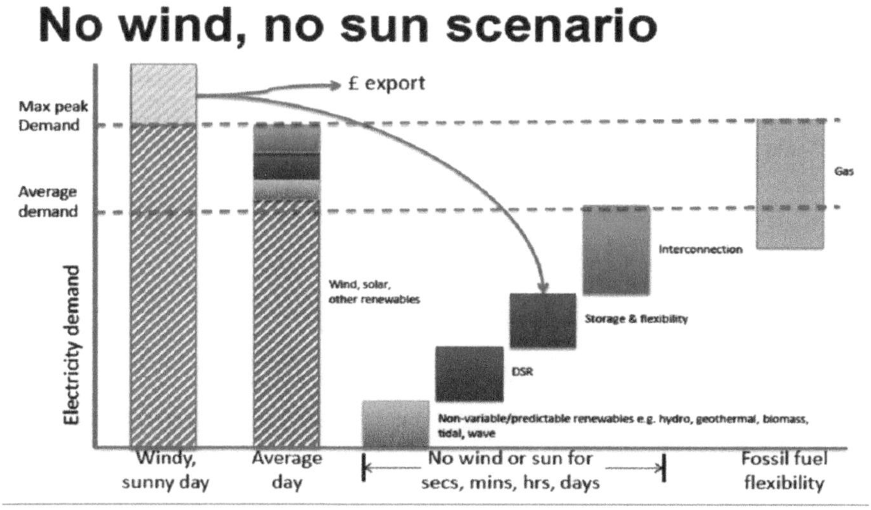

Figure 7.2. Balancing options and interactions. Reproduced with permission from Mitchell (2015).

7.2 Balancing technology issues

This book has tried to show that while no single grid balancing option will be sufficient on its own, taken together, in an integrated and widely interconnected approach, using a range of renewable sources, including the non-variable options, effective balancing should be possible. Certainly, the range of approaches is quite large. However, some may have problems.

We have already looked at some of the technical and economic issues, but there is also the question of social and environmental impacts. These are vital issues, given that the aim of balancing is to aid the development and use of clean, green energy systems. It is fortunate therefore that most of the balancing systems looked at in this book seem to be environmentally unproblematic, with perhaps the exception of pumped hydro storage. Some environmentalists have opposed large hydro projects, although that has mainly concerned major new projects in remote areas that might not be suited to pumped storage. Much of the current development work concerns modifying and perhaps linking up existing, often smaller, hydro projects, so that they can operate in pumped storage mode, although some new medium-scale systems are also being developed.

Some energy storage systems involve the use of toxic materials (e.g. in batteries), but as we have seen, non-toxic versions are being developed. Hydrogen storage and transmission has risks, but so does the storage and transmission of most fuels, with hydrogen arguably being one of the less risky options: it is lighter than air and so disperses easily. Supergrid links may be invasive, but as we have seen, there are options for overlaying them with existing grids, or putting some sections underground, although that would add to the cost.

The use of biomass as a firm source of energy, and for the production of biogas as a possible energy storage medium, opens up some environmental issues. Some environmentalists oppose the use of biomass for energy production on the basis of land use and ecological impacts: it depletes a crucial carbon sink, reduces the area available for food production and can undermine local biodiversity. However, biogas produced from farm and food wastes should not face these problems, and avoids the release of methane into the atmosphere, in which case it should be able to play a role in balancing variable renewables on a sustainable basis. Some other forms of biomass may also be less of a problem, short rotation coppicing of fast growing non-food crops for example, although clearly careful regulation is needed, as with all types of renewable system (Elliott 2015).

Although the environmental and safety issues do not seem insurmountable, there are some perhaps more challenging technical and operational issues to be resolved in relation to these systems, in addition to some economic market-related uncertainties. For example, in relation to interconnectors, it remains unclear how much excess output will be available when it is needed for trade. Much will depend on prices and market structures, as well on which renewables are developed and where they are located. Perversely, if supergrids spread, countries may be tempted to reduce their renewable overcapacity, and so have less to trade. It is also likely that there will be large surpluses in the summer, when demand is smaller and solar is at a maximum, during which time few trades will be viable or needed, with some of the renewable capacity being left idle. Storage capacity will also be unused then, unless it is inter-seasonal storage, and not much of that may exist.

Given that renewable capacity may have to be set higher than would otherwise be needed to meet demand, there will also be surplus production at other times, for example when wind output is high and demand low. As critics have pointed out, in some scenarios, assuming extensive renewable expansion, the surplus can be quite large and it would be very wasteful. However, as we have seen, this misses the point that not only can some curtailment be avoided by supergrid exports, but some of the excess can be stored directly or converted into valuable storable fuels for later use, possibly in other sectors (e.g. heating and transport).

Potential opportunities for cross-cutting support like this, along with the availability of other balancing options such as supergrids and smart grids, and reliable inputs from a range of non-variable renewables, suggest that it may in time be possible to limit or even avoid the need for fossil plant back-up, as well as the need for fossil fuel use generally. For many this is a key issue. It is one reason why some support renewables, and others do not. So it is worth setting out what seems to be the conclusion from the studies looked at in this book in that regard.

As we have seen, views differ. Some say that although variable renewables will have some capacity credit (perhaps, EDF says, around 20%, even in a 40% variable renewable scenario), extensive fossil plant back-up will still be required to deal with the occasional long lulls in wind and solar availability (Burtin and Silva 2015). Although otherwise confident that balancing measures can deal with renewable variations and reduce curtailment problems, even the strongly pro-renewables Agora Energiewende admits that 'situations can occur in which conventional power plants and imports must cover almost the entire load' (Agora 2015b). However, as we have seen, there is some evidence that this may not be the case, or at least that the need for fossil back-up may be limited. The next section reprises this issue, drawing on some of the UK studies looked at earlier.

7.3 Balancing renewables without fossil fuel use

In its review of sustainable energy UK pathways to 2050, the UK Department of Energy and Climate Change said that 'either significant storage, interconnection and other balancing technologies were likely to be required, or we would need to rely on extra back-up capacity' (DECC 2010).

As we have seen, the UK Energy Research Partnership did not think the non-fossil balancing options could help much, but, even so, calculated that with a hypothetical 100% renewable input from wind and PV solar meeting most electricity needs most of the time, only around 12% of total supply would have to be available from fossil plants (ERP 2015). That is in line with an earlier study by the consultant group Poyry, which we looked at in chapter 5. It found that with 94% of UK electricity being met from renewable sources by 2050, there would only be a need for 21 GW of new gas peaking plant, plus the residual 13 GW of other gas plants. That is 11.4% of the 298 GW of total capacity in their scenario, slightly less than the ERP's 12%, perhaps due to the wider range of renewables Poyry used (Poyry 2011).

This may seem surprising, but in most high renewables scenarios, there is usually a high level of overcapacity specified. For example, the Pyory scenario had a large amount of wind and other renewable capacity, 264 GW in all. Even if at times some of this might be unable to deliver much output, the capacity credit for the 189 GW of wind capacity would be around 13–14 GW. Along with the residual fossil input, this and the other renewable inputs should be more than enough—also assuming some imports, demand management and storage—to meet demand most of the time, given that around 60 GW is the usual UK maximum.

As we have seen this 'installing overcapacity' approach is one balancing strategy. As we have also seen, the downside is that this huge capacity would be much more than is needed most of the time, so there would often be a surplus (120 TWh per annum in the Poyry scenario) and the need for curtailment. Poyry did say that this surplus could be reduced if a different mix of renewables was used (their chosen mix was dominated by wind), but it would still be large.

However, again as we have seen, if this surplus could be used to make hydrogen to fuel standby gas plants, then, assuming some imports via interconnectors, along with other balancing services (storage and demand-side management), the need for a fossil fuel input to meet electricity demand peaks could be reduced substantially, perhaps to almost zero, and on a reliable basis. This was one of the conclusions from the high renewables scenario developed by the Pugwash group in the UK, with which I was involved (Pugwash 2013).

In terms of reliability, in the case of both the Poyry and Pugwash high renewables scenarios, the renewables mix was able to meet stringent reliability and energy security criteria. Poyry found that in its 'stretching but feasible' 94% renewable scenario, 'the electricity system was able to accommodate these high levels of renewable generation whilst complying with the specified constraints on emissions and security of supply'. It added that 'these scenarios were also tested against more extreme weather conditions, as defined as increased frequency of low wind periods ("lulls") and greater variability of wind output. In our (very) high renewable scenarios, we found that there is relatively little difference between the level of security of supply […] in an average weather year and from the level in one of our extreme weather years' (Poyry 2011).

The Pugwash 2050 scenario, aiming to meet around 80% of UK energy needs from renewables, backed up by some storage, demand management and interconnector imports, was run through the 2050 Pathways computer model developed by the UK Department of Energy and Climate Change, which included a 'stress test' that assumed five days of low renewable input and low temperatures. It passed easily, meeting electricity demand peaks successfully. As noted in chapter 5, it was also able to meet most demand for heat and transport fuels, with some heating and transport needs being met using renewable electricity. That input was also augmented by the direct use of renewables for heating (mainly solar and biomass) and for fueling vehicles (biogas and syngas), with, in its 80% renewables scenario, the residual fossil input taking up the slack, including a small input to electricity production.

However, it was suggested that this fossil residual could be eliminated by increasing the renewable capacity and using power-to-gas conversion of the occasional surplus to meet shortfalls. Similar results were obtained in the much more ambitious Zero Carbon Britain (ZCB) highly accelerated scenario developed by the Centre for Alternative Technology, which looked to achieve zero emissions by 2030, with renewables ramped up very rapidly (ZCB 2015).

That may be overambitious in deployment terms, but in both of these scenarios there were no problems in terms of electricity supply. It could also be balanced effectively, as could the energy needed for heat and vehicle fuels, since it can be stored easily. So the key remaining issue is whether there would be sufficient green energy sources available to cover these non-electrical needs without fossil inputs. That is unclear, especially for transport. The Pugwash study suggested that some of the surplus electricity generation, rather than being exported, could be used to make vehicle fuels, and certainly, as we have seen, some of the power-to-gas and power-to-liquid projects that have now sprung up look to that option.

Clearly, in ambitious scenarios like this (and the many others that have emerged since), transport is a major issue, as is the role of biomass in supplying the necessary energy. Some avoid biomass entirely (Delucchi and Jacobson 2013). The Pugwash 80% UK scenario did not include any imported biomass or biofuels, a policy supported by many environmentalists, but did include indigenous sustainable biomass production. If that is to expand significantly, it is usually argued that, given land-use constraints, food production and dietary patterns would have to change, as both the Pugwash and ZCB scenarios suggested.

This suggestion, along with the need to reduce demand significantly (by up to 40% in the Pugwash scenario), opens up wide social issues concerning the nature of future society and its consumption patterns, some of which I have explored in earlier books (Elliott 2003, Elliott 2015). However, as far as the more limited 'balancing' focus in this book is concerned, it is perhaps foolish to try to predict long-term technological developments, social changes and associated shifts in demand patterns.

That is not to say that long-range scenarios are of no use. They provide an opportunity to test out logical quantitative projections, priorities and assumptions in coherent frameworks. Certainly, some of the more quantitative scenarios provide useful guides to balancing requirements, and in some cases follow that through in some detail, developing some helpful analysis techniques (Sorensen 2014, Jacobson *et al* 2015a, Jacobson *et al* 2015b). Going much further than that, for example to detailed costings, is, at present, probably not sensible or realistic, as noted above, given the uncertainties about what may actually be possible and needed in the long term. We can nevertheless perhaps usefully address possible reactions to, and deployment problems with, some of the proposed balancing systems, since they are likely to start impacting in the medium-term future and may influence the pattern of development. In the final section below we look at some of these issues and at some of the other wider challenges that are likely to influence the deployment and choice of systems.

7.4 The challenges ahead

In addition to the specific technical, economic, safety and environmental issues looked at in this book, there is also a wider range of social, political and strategic issues to consider when deciding on the optimal mix of supply and balancing systems, including public reactions to deployment.

Certainly, public reactions to the balancing systems will be important, as they have been in the case of renewable deployment generally. On the supply side, some of the changes will essentially be invisible to the public. Most utility-run storage facilities, pumped hydro apart, will look much like any other plants, while the same can be said for most back-up plants and associated biogas/syngas stores and power-to-gas conversion plants. As indicated above, supergrid links may, however, be much less invisible, although, like hydro, they would be remote from most people and with supergrids there are options for going underground in sensitive areas.

On the demand side, there would be more intimate and widespread interactions. It is not clear how consumers will react to smart grid demand management and time-of-use pricing systems, and how much they will help to limit peaks and avoid energy waste. Most consumers will probably not want to have to be bothered with energy management, so automated systems may prove acceptable, as long as consumers see some benefits. These benefits, however, may be longer term and not just economic.

However, if the new system can be seen to deliver energy reliably, without undue costs or impacts, and without too much need for behavioural change, then it might be widely supported. Then again, some consumers seem to be willing to voluntarily adopt new energy systems, perhaps wishing to make personal contributions to energy sustainability on behalf of future generations. The growth of the 'prosumer' self-generation movement in many ways challenges the technological and market status quo, making old certainties redundant: a new system may be emerging. That will open up new social and political issues (Heinberg 2015b, Sholten and Bosman 2016, Koiralaa *et al* 2016).

While some of the economic problems of the new system may look significant, for example in relation to short-term cost effectiveness and system impacts, that may be too narrow a framework, as, amongst many others, Sorensen has argued (Sorensen 2014). All systems have their problems, especially at an early stage, while some apparently successful systems begin to break down as the context changes. We have become used to an energy system in which large centralised plants deliver energy via grids to large numbers of distributed consumers. That worked reasonably well in supply terms, but was actually very inefficient in energy terms. That did not matter when energy sources were abundant and relatively cheap and could be used without regard to their environmental impacts. We are no longer in that situation. The newly emerging more decentralised system and the associated balancing arrangements described in this book will have shortcomings and problems, but there seems to be no easy alternative, unless we are willing to try for one last spin of the centralisation approach based on some form of nuclear technology.

This is not the place to revisit all the economic, safety and security problems of the nuclear option, or the issue of the sustainability of fissile fuel supplies. Suffice it to say that most countries are not going down that road, while some that did initially have backed off, most notably, post-Fukushima, Germany, but also Belgium, Italy and Switzerland (Elliott 2013). Even France is now cutting back drastically and despite plans for replacements, nuclear seems to be stalled in the USA, with many old plants being closed early. Japan, of course, shut all its nuclear plants after the Fukushima disaster, although a few have now been restarted, despite much opposition. No new plants are likely to be built there. It is true that some countries are still pushing ahead with large nuclear expansion programmes, notably the UK, Russia, India and China, but some of these programmes seem to be facing economic problems: while renewables are getting cheaper, new nuclear seems to be getting more expensive, at least in the EU and US. Even in China, where costs are evidently lower, renewables already produce around ten times more electricity than nuclear, with wind output alone having overtaken that from nuclear. Globally, the nuclear contribution to electricity supply seems to be static, at around 11%, with the renewable expansion outpacing it, having already reached over 22% (WNISR 2015, REN21 2015).

As ever, there are promises that new nuclear technology will be safer and cheaper and better matched to needs, perhaps through moving to smaller scales. In the past, attempts were made to make nuclear more economic by going for larger plants. That has not been very successful. It is not clear that going to smaller plants will be any more successful, or that it will avoid safety and security risks (Mackerron 2015).

Absent that option, or some new breakthrough, nuclear power seems unlikely to prosper. It may co-exist for a while, and remain important in some countries, but in general, as with other large-scale centralised base-load energy systems, it seems to have no place in the emerging more flexible, decentralised systems.

Views like this used to be the province of fringe environmentalists. But now we are hearing the same from major companies like Siemens, RWE and E.ON in Germany, and some of their counterparts elsewhere. For example, Steve Holliday, chief executive officer of the UK's National Grid, says that the idea of large coal-fired or nuclear power stations for base-load power is obsolete, as energy markets move towards much more flexible distributed production and smart grids: 'The idea of base-load power is already outdated. The future will be much more driven by availability of supply; by demand side response and management' (Holliday 2015).

Certainly, on the supply side, many detailed scenarios have now been produced in which renewables supply the bulk of electricity, 80–90% or more, in many countries by around 2050, assuming continued improvements in energy efficiency, and some also have renewables meeting most heat and transport demand (ECF 2010, EREC 2010, PWC 2010, WWF 2013, WWF 2014). Indeed, some scenarios suggest that getting near 100% of all energy from renewables is possible, globally, by 2050 (Delucchi and Jacobson 2011, Greenpeace 2015). That may be optimistic, but it does seem to be the right direction of travel, and a good, if challenging, target (Elliott 2015).

It will not be easy to make this type of transition, but hopefully this book will have provided a guide to what issues may lie ahead in terms of grid balancing and system integration. There may be practical limits to how much balancing can be provided by each of the technical options, but as the International Energy Agency put it in an early review 'there is no intrinsic, technical ceiling to variable renewables' potential. Variability has to be looked at in the context of power system flexibility: if a power system is sufficiently flexible, in terms of power production, load management, interconnection and storage, the importance of the variability aspect is reduced' (IEA 2008).

Experience and developments since that was written seem to suggest that, as this book has hopefully indicated, balancing variable renewables may not be a major constraint on the expansion of the use of renewables.

Moreover, this need not be expensive. Indeed, a recent study by the UK government's National Infrastructure Commission, looking to a low carbon future, has suggested that the development of smart grid demand response, energy storage and interconnectors, so as to better balance energy supply and demand, could save consumers up to £8bn p.a. by 2030 (NIC 2016).

Interest in these new balancing technologies is spreading across the world, with, as recent reviews indicate, valuable experience being gained as they are introduced as part of the energy system transformation process (Miller *et al* 2015, Martinot 2016). A key requirement for a successful transition to a system based increasingly on renewable energy is thus being developed. There will clearly be challenges ahead, but as the costs of inaction and the benefits of change become more apparent, the pace of change is likely to accelerate.

Chapter summary

1. An energy system change process is underway and is unlikely to be halted, given the need to respond to the environmental impacts of the existing system.
2. There are many technical, economic and institutional challenges facing the grid-balancing options and there is a need for policy change.
3. Despite the challenges, it does seem possible that renewables can provide a reliable basis for a sustainable energy future, with low impacts and costs.

References

Agora 2015a *The Integration Costs of Wind and Solar Power* (Berlin: Agora Energiewende) www.agora-energiewende.de/fileadmin/Projekte/2014/integrationskosten-wind-pv/Agora_Integration_Cost_Wind_PV_web.pdf

Agora 2015b The European power system in 2030—flexibility challenges and integration benefits *Fraunhofer Institute Report for Agora Energiewende* www.agora-energiewende.org/service/publikationen/publikation/pub-action/show/pub-title/the-european-power-system-in-2030-flexibility-challenges-and-integration-benefits/

Bogdnov D and Breyer C 2015 Eurasian super grid for 100% renewable energy power supply: generation and storage technologies in the cost optimal mix ISES Solar World Congress

2015 (Daegu) www.researchgate.net/publication/283713531_Eurasian_Super_Grid_for_100_Renewable_Energy_power_supply_Generation_and_storage_technologies_in_the_cost_optimal_mix

Burtin A and Silva V 2015 Technical and economic analysis of the European electricity system with 60% RES *EDF R&D* www.energypost.eu/wp-content/uploads/2015/06/EDF-study-for-download-on-EP.pdf

DECC 2010 *2050 Pathways Analysis* (London: Department of Energy and Climate Change) www.decc.gov.uk/en/content/cms/what_we_do/lc_uk/2050/2050.aspx

Delucchi M and Jacobson M 2011 Providing all global energy with wind, water, and solar power. Part II. Reliability, system and transmission costs, and policies *Energy Policy* **39** 1170–90

Delucchi M and Jacobson M 2013 Meeting the world's energy needs entirely with wind, water, and solar power *Bull. At. Sci.* **69** 30–40

ECF 2010 *Roadmap 2050* (Brussels: European Climate Foundation) www.roadmap2050.eu

EREC 2010 *Rethinking 2050* (Brussels: European Renewable Energy Council) www.rethinking2050.eu

Elliott D 2013 *Fukushima: Impacts and Implications* (Basingstoke: Palgrave Pivot)

Elliott D 2003 *Energy, Society and Environment* (London: Routledge)

Elliott D 2015 *Green Energy Futures* (London: Palgrave Pivot)

Greenpeace 2015 *Energy[R]evolution* 5th edn (Greenpeace/Global Wind Energy Council) www.greenpeace.org/international/Global/international/publications/climate/2015/Energy-Revolution-2015-Full.pdf

Heinberg R 2015a *Our Renewable Future* (Santa Rosa CA: Post Carbon Institute) www.postcarbon.org/our-renewable-future-essay

Heinberg R 2015b *Renewable Energy after COP 21* (Santa Rosa CA: Post Carbon Institute) www.postcarbon.org/renewable-energy-after-cop21/

Holliday S 2015 'Base-load is outdated'—National Grid CEO says *Energy Post* www.energypost.eu/interview-steve-holliday-ceo-national-grid-idea-large-power-stations-baseload-power-outdated/

IEA 2008 *Variable Renewables Options for Flexible Electricity Systems* (Paris: International Energy Agency) www.iea.org/publications/freepublications/publication/Empowering_Variable_Renewables.pdf

Imperial/NERA 2015 Value of flexibility in a decarbonised grid and system externalities of low-carbon generation technologies *Imperial College London/NERA Consultants Report for the Committee on Climate Change* www.theccc.org.uk/publication/value-of-flexibility-in-a-decarbonised-grid-and-system-externalities-of-low-carbon-generation-technologies/

IRENA 2014 *Renewable Power Generation Costs in 2014* (Abu Dhabi: International Renewable Energy Agency) www.irena.org/menu/index.aspx?mnu=Subcat&PriMenuID=36&CatID=141&SubcatID=494

Jacobson M, Delucchi M, Cameron M and Frew B 2015a A low-cost solution to the grid reliability problem with 100% penetration of intermittent wind, water, and solar for all purposes *Proc. Natl Acad. Sci.* **112** 15060–5

Jacobson M *et al* 2015b 100% clean and renewable wind, water, and sunlight (WWS) all-sector energy roadmaps for 139 countries of the world (Stanford, CA: Stanford University) http://web.stanford.edu/group/efmh/jacobson/Articles/I/CountriesWWS.pdf

Koiralaa B, Kolioua E, Friegec J, Hakvoorta R and Herdera P 2016 Energetic communities for community energy: a review of key issues and trends shaping integrated community energy systems *Renew. Sustain. Energy Rev.* **56** 722–44

Mackerron G 2015 *Small Modular Reactors—a Real Prospect?* (Sussex Energy Group, University of Sussex) http://blogs.sussex.ac.uk/sussexenergygroup/2015/10/09/small-modular-reactors-a-real-prospect-by-gordon-mackerron

Martinot E 2016 Grid integration of renewable energy: flexibility, innovation, experience *Annu. Rev. Environ. Resour.* http://martinot.info/Martinot_AR2016_grid_integration_prepub.pdf

Miller M *et al* 2015 *Status Report on Power System Transformation* (Golden, CO: US National Renewable Energy Laboratory) www.nrel.gov/docs/fy15osti/63366.pdf

Mitchell C 2015 A 100% renewable energy system operation on no wind, no sun days *IGov, Exeter University* http://projects.exeter.ac.uk/igov/no-resource-is-100-reliable-a-100-renewable-energy-system-operation-on-no-wind-no-sun-days/

NIC 2016 *Smart Power* National Infrastructure Commission, London www.gov.uk/government/uploads/system/uploads/attachment_data/file/505218/IC_Energy_Report_web.pdf

NOAA 2016 Rapid, affordable energy transformation possible *Press Release* (25 January 2016) (Washington, DC: National Oceanic and Atmospheric Administration, US Department of Commerce) www.noaanews.noaa.gov/stories2016/012516-rapid-affordable-energy-transformation-possible.html

Pöyry 2011 Analysing technical constraints on renewable generation to 2050 *Report for the Committee on Climate Change* www.theccc.org.uk/archive/aws/Renewables%20Review/232_Report_Analysing%20the%20technical%20constraints%20on%20renewable%20generation_v8_0.pdf

Pugwash 2013 *Pathways to 2050: Three Possible UK Energy Strategies* (London: British Pugwash) http://britishpugwash.org/pathways-to-2050-three-possible-uk-energy-strategies/

PWC 2010 *A Roadmap to 2050 for Europe and North Africa* (London: PriceWaterhouse Coopers) www.pwc.co.uk/eng/publications/100_percent_renewable_electricity.html

REN21 2015 2015 *Global Status Report* (Renewable Energy Network for the 21st Century) www.ren21.net/status-of-renewables/global-status-report/

Scholten D and Bosman R 2016 The geopolitics of renewables; exploring the political implications of renewable energy systems *Technol. Forecast. Soc. Change* **103** 273–83

Sorensen B 2014 *Energy Intermittency* (London: Routledge)

WNISR 2015 *World Nuclear Industry Status Report* (Independent Annual Review) www.worldnuclearreport.org/

WWF 2013 *100% Renewable Energy by 2050 for India* (New Delhi: World Wildlife Fund for Nature/TERI) www.wwfindia.org/?10261/100-Renewable-Energy-by-2050-for-India

WWF 2014 *China's Future Generation* (World Wildlife Fund for Nature) http://worldwildlife.org/publications/china-s-future-generation-assessing-the-maximum-potential-for-renewable-power-sources-in-china-to-2050

ZCB 2015 *Zero Carbon Britain* (Machynlleth: Centre for Alternative Technology) www.zerocarbonbritain.com

Lightning Source UK Ltd.
Milton Keynes UK
UKOW07n1051150516

274238UK00002B/6/P